Honey:
From source to sale and show bench

Jeff Rounce

NORTHERN BEE BOOKS

© Jeff Rounce
First published 1990 by Northern Bee Books,
Scout Bottom Farm, Mytholmroyd, Hebden Bridge.

Cover illustration Richard Alston

British Library Cataloguing in Publication Data
Rounce, Jeff
        Honey : from source to sale & showbench.
        1.   Honey. Manufacture
        I.   Title
        638.16

        ISBN 0-907908-49-7

Printed and bound by
Arc & Throstle Press, Todmorden.

Contents

Ilustrations

Forward: Honey

In these days when we are more conscious than ever before about the purity of our food, it is essential that beekeepers should be aware of the properties of the product they sell or exhibit. Much mystique still surrounds all aspects of beekeeping and hive products and the perpetuation of some of the myths does little to enhance our craft.

Honey in all its variety of colours, aromas, flavours, consistency and other properties is an important food item in that it does not have to be digested, most of its complex molecules have been reduced to a form that can be assimilated without the need for digestion. This makes honey a very useful, readily available form of energy to athletes and other sportsmen, and also to invalids for whom the process of digestion would be an additional burden on an already weakened body.

The comments and questions from spectators at honey shows and other events where honey is being exhibited or sold reveals how much we as beekeepers need to know in order to satisfy the various enquiries that are made, and to dispel errroneous ideas. Most people are intensely interested in the origins of honey, the reasons for different shades, and the difference between liquid and granulated honey. Many still believe that bees collect pollen and convert it into nectar; while some think that bees collect honey. Full credit is rarely given to the work bees do by pollinating when collecting nectar. Comments regarding the uses of honey as a medicine or aphrodisiac are not uncommon. Visitors to honey shows leave happier if they can be told why the judge has given the awards to certain entries while others stand naked on the showbench. Gone, I hope, are the days when honey judges list themselves among the untouchables and I take my hat off to those judges who, if the public are present during judging, pause in their arduous task to answer questions sometimes posed, or stay behind after the judging is finished and the show cards displayed, to discuss the awards or lack of them with the exhibitors.

I am appalled, not only as an exhibitor, but also as a judge at the low standard of some entries and can only hope that honey offered for sale by such exhibitors is of a higher standard. I applaud those show officials who, if honey, is being offered for sale, refuse any honey which, is in the judge's opinion, not of a sufficiently high standard. If we, as producers of honey, are to compete with usually much cheaper imported honey, we can only do so if in all respects, quality, handling and appearance, our product can be shown to be superior.

In writing this book I am aware that apart from my personal opinions or methods of working, it contains little that is original. What is probably original, and this was the publisher's idea not mine so I take no credit for it, is the idea of putting all aspects of honey into one book.

1

This has been done in the past for beeswax, skeps, queen raising, disease, bee hives and many other aspects of beekeeping but not to my knowledge to cover the history, sources, extracting, handling, marketing, costing, showing and judging of honey. I hope it will not be treated as just another beekeeping book written by someone hoping to make a name for himself. I have not been able to include all the wealth of material which is available to those interested in honey, so a list of those books which I have found most useful to me in the past and in compiling this book are listed at the end. I unashamedly admit my debt to the authors.

My thanks must also go to various other authors who have unwittingly helped me, also to my various friends and aquaintances, many sadly no longer with us, who over the past 40 odd years have given help, advice, criticism and much more the sort of friendship that can only be found among beekeepers; they include all from the internationally well known to the obscure and sometimes almost inarticulate. Especially I thank my long time friend and erstwhile country beekeeping adviser to Norfolk, Paul Metcalf, for reading the manuscript, together with his comments, criticism and advice.

Finally, to my wife for her unfailing help in washing thousands of honey jars over the years, enduring wax and honey over floor and furniture in amounts no normal person would endure; for the ungodly hours she has spent helping to move beehives and the considerable discomfort she has endured in the process.

Chapter 1:
Introduction

Man's relationship with bees goes back into the mists of time; originally that relationship could have differed little from that of other mammals who appeared much earlier on this earth, that is the robbing of bee colonies for honey and brood both of which would be avidly consumed on the spot or taken back to the rest of the tribe for immediate consumption. In other words bee-hunting. It probably became a less unpleasant activity when man discovered that he could 'control' bees by the use of smoke.

From ancient writings in the Old Testament and other literature of pre-Christian times, from cave and and wall paintings in Africa, Spain and India, and from bas-relief of Ancient Egypt, it is obvious that man's interest in bees has been a lengthy one.   The use of the bee as a hieroglyph in ancient Egyptian picture writing gives an idea of the bee's importance six millennia ago.

The mere fact that something so small and insignificant as an insect could produce a substance of such exquisite sweetness, as well as beeswax, gave it an air of mystery so that in addition to its use as a sweetner, honey became endowed with magical properties, not only had it healing powers, was believed to be an aphrodisiac, but it also had a magico-religious significance. Not for nothing are nuptial celebrations known as the honeymoon. Even in today's sophisticated society there still exists a mystique about bees, honey and some other hive products.

Over the centuries beeswax has been used for candles, embalming, writing tablets, cosmetics, polish, encaustic painting, and lost wax casting among others, while apart from the use of honey as a sweetener it is the basis of mead, has been used to assuage the gods, as a valuable gift to important persons and in religious ceremonies. So important did it become that it is hardly surprising that man attempted to make it another domestic animal alongside cattle, sheep, goats, horses and others, not to mention the cat and dog. How the transition from bee-hunting to beekeeping took place can only be conjectured. It may be those in forested areas where bees lived in hollow trees, these were felled, cut into sections and that containing the colonies of bees taken home. It is more likely that upon seeing that swarms of bees would occupy any hollow place which provided them with shelter, space to build the combs and an access large enough for entry and exit but small enough to defend, that our forefathers set up such devices in the hope that these would be occupied by swarms. Such devices might have been made of clay, bark, woven straw or whatever material was suitable and readily available. Straw skeps, bark hives and clay pots are still in use and, until recently, so were New Zealand butter boxes, in Ireland. Rev. L. Langstroth is reputed to have taken the dimension for his hive from the boxes in which champagne was imported into America. Those of you who have been called upon to remove swarms will testify to the peculiar cavities occupied by swarming bees. I had once to remove such a colony from a five gallon oil drum which had been casually thrown away, but not far enough from the house for the occupant to be happy.

It would appear that it was mainly in those areas that enjoyed a relatively moisture free air, in particular, Egypt, that honey could be kept without fermenting for any length of time. It was probably from attempts at storing honey that mead making had its beginnings, also no coincidence that mead became a national drink in countries of northern Europe where honey was difficult to store without it fermenting. As Europeans set out to colonise and settle in all other habitable areas of the planet, they found those countries where honey bees did not exist and where beekeeping was unknown. Eventually European honey bees were taken to North and South America, Australia and New Zealand to name but a few parts of the world where honeybees now thrive and from which most of the honey which is eaten in Europe now comes. In more recent times apiculture has developed to be an important industry in countries such as Mexico and China.

With the introduction of bee hives with moveable frames, the queen excluder and centrifugal extractors, came the transition from honey in the comb to honey in the jar, the latter becoming more popular as legislation was brought into ensure the purity of honey and the penalisation of the practice of adulteration with other cheaper sweeteners, a practice still unfortunately practiced today, but much more easy to detect. All these have resulted in honey becoming an important food in its own right and as a sweetener in cooking and confectionery. It is still used medicinally and as an animal foodstuff, considerable quantities being fed to racehorses. About 95% of the world's honey production is marketed as honey for human consumption. With the exception of central Africa, beekeeping worldwide is devoted to honey production often with crop pollination as a sideline, only in Africa is beeswax the main product.

Of the 20,000 species of bees, both solitary and social, there are only four species that can be called honey bees, although many species of social bees e.g. Bumble and stingless bees of the sub-families Bombinea and Meliponini respectively, store small quantities of honey. The four species of honey bees, *Apis*, are *mellifera, cerana, dorsata* and *florea*. Originally they were native to Europe, Africa and Asia, being absent from Australasia, east of the Wallace Line, and

America.

Apis mellifera was native to Europe, Africa and the Middle East as far as south west from where its range slightly overlaps that of A. florea. Apis cerana, which is only about half the size of A. mellifera, has a range extending from the Persian Gulf through the Indian sub-continent into south-east Asia and Indonesia as far as the Phillipines, Borneo and Java, it extends northwards up east Asia to Japan, China and Korea. Both A. mellifera and A. cerana 'nest' in protective cavities and can adapt to living in hives.

The largest honey bee, A. dorsata, is found only in the Indian sub-continent and south east Asia, while the smallest honey bee, A. florea, has a similar range eastwards, its range extends westwards as far as Oman. Both these species of honey bees build single combs in the open, usually under the branches of trees but also under the eaves of houses and over-hangs on vertical rock faces. Neither will adapt to living in hives and honey is obtained by robbing the colonies, whereas A. mellifera and cerana can be encouraged to store honey for extraction in parts of the hive, the supers, from where it can be removed without much interference with the colony concerned.

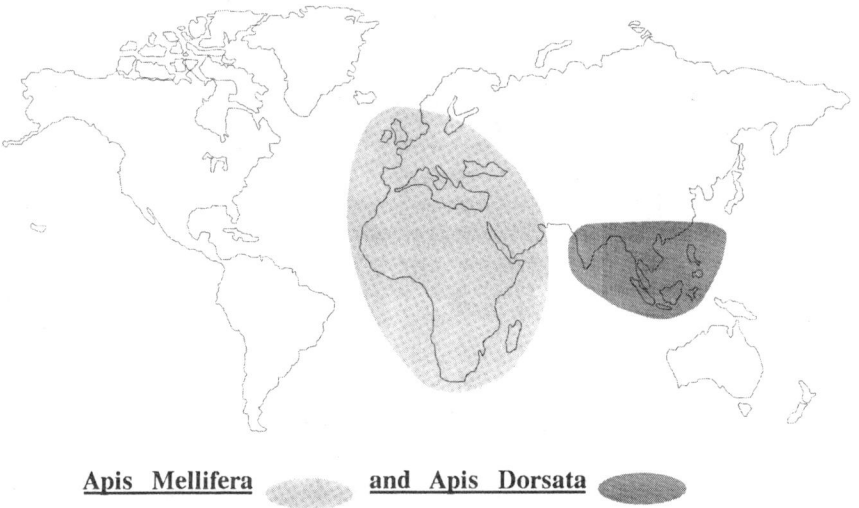

Apis Mellifera ⬮⬮⬮⬮ **and Apis Dorsata** ⬮⬮⬮⬮

Fig. 1 Maps showing areas in which honey bees are indigenous.

Apis Cerana

Apis Florea

Man has increased the range of *A. mellifera* as already mentioned and as a result of introduction into some parts of south east Asia, the parasite, *Varroa Jacobsoni*, originally a parasite of *A. cerana*, has become widespread in colonies of *A. mellifera* in other parts of Asia, most of Europe and Africa, and also America both north and south.

More recently *A. florea* has been found in an area around Khartoum airport in the Sudan, where it appears to have been introduced from Pakistan. It has to be assumed that a stray swarm was carried to that country by aircraft.

Chapter 2:
What is Honey?

It would be wrong to describe honey as processed nectar, for honeybees will collect and process any sweet substance. To be precise, it is a supersaturated solution of sugars and other substances in water; as such some of the sugar comes out of the solution in the form of crystals and granulated or 'set' honey results. Its high sugar/low water content together with its acidity ensures that provided the proportions of the constituent parts remain unchanged it will not ferment as many other sugar solutions do. Being hygroscopic however, it will, if conditions permit, absorb moisture from the air and the yeasts present in the honey will cause fermentation and mead will result, if the relative humidity of the air is higher than that of the honey.

Fig. 2

Extra-floral nectaries on braccken (Pteridium aquilinum): small, often coloured swellings at junctions of branching stalk of the frond.
[After Pontonié in Bower, F.O. "The Ferns", C.U.P., 1923]
Drawn by A.E. Calder

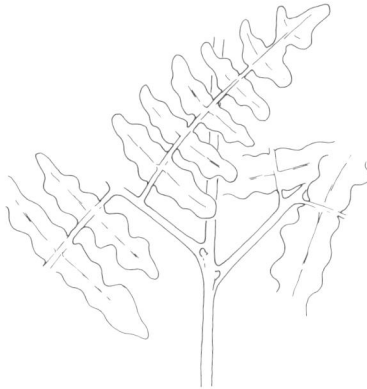

During the Cretaccous Period which commenced about 130,000,000 years ago, angiosperms, the flowering plants, evolved from fern like ancestors. With this evolution came the need for pollen produced by the stamens of the andrecium to be transferred to the pistil of the gynaecium otherwise the flower would not be pollinated and subsequent fertilisation and seed production would not take place. Pollination became an essential part of the reproduction of the major group of plants on this planet.

Flowering plants have developed a variety of methods of pollination, a few water plants use water, while many have evolved to use wind. A very large number and they are those that bear very obvious flowers have evolved to use animals as the agents for pollen transfer. The

8

animals may be small mammals or small birds but since insects form the largest single group of animals and the majority are winged, it is hardly surprising they are the main agents for pollen transfer other than wind. Insect pollinated flowers are termed *entomophilous*, wind pollinated *anemophilous*.

In order to be obvious and attractive to insects, two developments took place which are not to be found in wind or water pollinated flowers, the first was the evolution of large and brightly coloured petals, never green or black, and the production of nectar as a bait. To direct insects into the centre of the flower a pattern of lines are present on the petals, the so called nectar guides. These are not very obvious to the human eye unless illuminated with ultra violet light but since bees can see in that part of the electro-magnetic spectrum they are guided to the centre of the flower where the sexual organs and nectaries are placed. In so doing an insect will touch both stamens and sticky pistil. It is interesting that some plants have developed extra floral which can play no part in pollination, thus nectaries are often found on the midribs of leaves or leaf stalks as with the Cherry Laurel (*Prunus Laurocerasus*) and Wild Cherry (*Prunus cerasus*) respectively; they are also found on the stipules of the flower stalks of the Broad/Field Bean (*Vicia faba*). The most unusual extra floral nectaries are those found on a non-flowering plant, the fern Bracken (*Pteriddum acquilinum*).

Many flowers have evolved intricate structures or methods of timing the dihiscence of the anthers or opening of the lobed stigma together with the positioning of stamens and stigma, to ensure cross pollination. This results in nectar secretion from each flowers covering several days. Some plants, particularly trees bearing catkins, have separate male and female flowers while trees such as the Goat/Pussy Willow (*Salix caprea*) and other willows, also the Holly (*Ilex aquifolium*) are dioecious, they have male and female flowers on separate trees. The study of flower structure and plant-insect relationship is a study on its own and mentioned solely to show the intimate relationships that have evolved for the mutual benefit of both. Many insects have become totally dependent on flowers for all their nutritional requirements, this includes all the *Apoidea* of which there are about 20,000 species and of which the *Apidea*, the social bees is only a small group.

Nectar, which is processed from phloem sap by the nectaries, is a solution of sugars and small amounts of other substances, in water. The amount of water can vary from around 40% to 90% depending on a number of variable factors such as the species of plant producing it, weather conditions, particularly temperature and humidity, the availability of soil water and soil type. Honey bees appear not to be attracted to nectar having a sugar content of less than 20% although they may work the flowers for pollen. Nectar is not secreted continuously but normally at certain periods of the day so that foraging bees may transfer their foraging activities according to availability of nectar and competition from flowers with a higher sugar content in their nectar.

At one time it was thought that cane sugar, sucrose was the major constituent of the sugars in nectar, but now that a precise analysis of sugars is possible, it is known that there are three basic groups of nectar depending on which sugar predominates; either sucrose or its breakdown products glucose (*dextrose*) or fructose (*laevulose*). Since a molecule of sucrose is broken down into a molecule of glucose and one of fructose, if sucrose was the main sugar in nectar we would expect the characteristics of all nectars to be similar, particularly their sweetness and rapidity of granulation; this of course is not so. The three nectar groups are those in which sucrose predominates, nectars in which there are roughly equal properties of sucrose, glucose and fructose, and those in which glucose and fructose are in roughly equal proportions. Particular types of flowers or species of plants produce a particular type of nectar, this becomes obvious when one considers the characteristics of the honey produced by certain plants.

The conversion of nectar to honey involves two distinct processes both of which occur almost simultaneously, one is the evaporation of water to reduce the water content to between about 17 and 20%, the other is the reduction of the sucrose to glucose and fructose. The first process takes place in the hive and the second as soon as the bee has swallowed the nectar. At one time it was thought that bees could remove some water from the nectar while it was in the proventriculus - honey crop, in which it is carried back to the hive. Careful analysis of the water content of the nectar before bees collect it and when they regurgitate it back at the hive, has shown that there is a slight increase in water, probably due to added enzymes etc. from the hypopharyngeal glands.

The main enzyme present in the secretion from the hypopharyngeal glands is invertase, there are also small amounts of other enzymes, diastase which hydrolises starch is one of them. These enzymes are added to the nectar as it is swallowed and also by housekeeping bees while in the process of reducing the water content: like most, enzymes, invertase has an optimum temperature range for its activityy, this seems to be in the range 35° to 40°C.

When foraging bees return to the hive with a load of nectar the contents of the honey crop are regurgitated and given to housekeeping bees which are usually more than six days old, the majority of such bees are aged about 12 days old. The bees swallow and regurgitate the nectar repeatedly, exposing it to the currents of warm air in the hive as a flattened drop in the bend of the partly extended tongue. Repeated swallowing is probably done to add more enzymes to the nectar. When satisfied that they can process the nectar no further, the bees place it in an empty cell in the brood nest, the warmest part of the hive where the temperature approximates 95°F/31°C. At this temperature conversion of disaccharide sugars to glucose and fructose proceeds rapidly and much water is lost by evaporation to the warm air which is being circulated by fanning bees. The process of ripening, a term given by beekeepers to the conversion of nectar to honey, is completed in honey storage cells above the brood nest; these cells form an arc in the top of the brood frames or in the supers. When almost filled with honey, the cells are sealed with an impervious layer of beeswax to prevent atmospheric moisture diluting the honey and causing fermentation. At the end of a nectar flow, sometimes erroneously called a honey flow and also at the end of summer when foraging has ceased, some cells may not be filled sufficiently for the bees to seal them.

Honey varies considerably in its component parts and a list of various honeys and the amount of each component would not only be tedious for the reader but unnecessary to readers of this book; a list of books for further reading is given at the end for those who wish to pursue the subject further. The list of substances present in honey and given by me must be taken as 'average' honey; only when it is necessary to describe a particular type of honey for a specific reason will greater details be given.

Taking honey with 17% moisture content, 68% will be a mixture of the two sugars glucose and fructose. Incidentally, in addition to being known by their alternate names dextrose and laevulose respectively, they are also known as grape and fruit sugar. Unconverted sucrose makes up about 2% and the remainder consists of dextrin, proteins, minerals, acids and other substances in very small quantities, but very important, in that they produce the subtle aromas and flavours of different honeys.

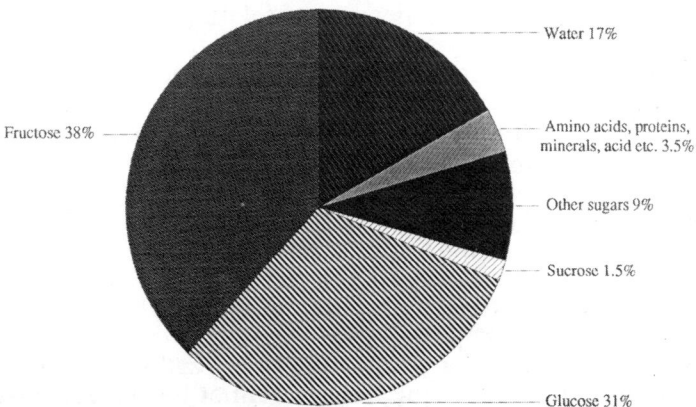

Fig. 3 Average composition of honey

The proportions of glucose and fructose are important in that they determine how rapidly any particular honey granulates. It is the glucose in honey that granulates, as it does so the water in which it was previously dissolved is added to that in which the fructose is dissolved thus raising the percentage of water present, as a result fermentation could take place in the honey when granulated, whereas it could not in the same honey when liquid. Because glucose crystals are white, granulated honey is always lighter in colour than it was when liquid. Those honeys which have a high level of fructose in relation to glucose granulate very slowly sometimes taking years to do so. Typical examples of very slow granulating honeys are False Acaria/Black Locust (*Robinia pseudoacacia*), White Clover (*Trifolium repens*) and Yellow Box (*Eucalyptus meliodora*). All these have roughly 40% fructose and 28% glucose.

On the other hand honeys in which the levels of glucose approximate or even exceed those of fructose granulate exceedingly rapidly, often on the combs. The best known of these are the *brassica* honeys especially those from the two species of oilseed rape, *brasscia rapus* and *brassica compestris*, whose honey must be extracted as soon as it is ripe otherwise it becomes impossible to extract it by normal methods. Other cruciferous plants such as the two mustards Black Mustard (*Brasscia nigra*) and White Mustard (*Synapsis alba*) also produce honeys which granulate fairly rapidly, but not as rapidly as oilseed rape, such honey must be extracted soon after the crops have finished flowering.

The rate of granulation determines the crystal size, this is true not only of sugar but of any substance which crystallises: salt, igneous rocks, etc. Rapid granulation results in the formation of small crystals, so rapidly granulating honeys are characterised by their smooth texture, on the other hand slowly granulating honeys have large, coarse crystals and usually set hard. Such honeys do not find favour with customers as they do not see the funny side of spoons bending in the middle when trying to remove the honey from the jar, or having removed it, find it is impossible to spread. For these reasons oilseed rape honey is the ideal substance for seeding honey when producing creamed honey, especially if the honey is one which is likely to granulate slowly and coarsely.

Heather honey is peculiar among honeys in that it is one of the few that is thixotropic; other honeys of this nature come from New Zealand (*Manuka*) and India (*Karvi*) so they are not likely to be encountered in the British Isles. Thixotropic honey is jelly-like when undisturbed but becomes a vicious liquid for a time after being agitated, allowing time for it to be bottled. Thixotropic honeys cannot be extracted from the combs without treatment; for this reason much heather honey is sold as comb honey. The term heather honey is a misnomer as it is produced not from heather (*Erica sp.*) but from a closely related plant Ling (*Caluna vulgaris*). Its thixotophy is due to the amount of protein colloids present, up to 1.8% as against 0.2% in other honeys.

Honey has a pH of around 4 which means it is fairly acid (a pH of 7 is neutral, neither acid nor alkaline). It also has a specific gravity of about 1.4, this means that for any given volume honey is 1.4 times as heavy as water. The specific gravity of honey varies with both water content and temperature, it decreases with an increase in water content and temperature. The term density is used to indicate the weight of a given volume of any substance, it may be grammes per cubic centimetre, pounds per cubic foot, kilogrammes per cubic metre, etc. Honey having a specific gravity of 1.4 has a density of 1.4 grammes per cubic centimeter. Confusing? It shouldn't be.

The density of honey also determines the viscosity, the higher the density, the greater the viscosity; thus viscosity is related to water content. Both water content and viscosity are important factors in showing and judging honey as well as producing honey for the market.

Two other components of honey are of importance as they are indicative of the age of any honey as well as mishandling during the processes that take place between extraction and bottling. The amounts of these substances present in honey are governed by E.E.C. regulations. The first of these is the enzyme diastase, also called amylase. Because enzymes are proteins they are degraded with high temperatures, especially those above 140°F/60°C. Its presence is measured in arbitrary units which denote the level of diastase activity, this decreases with time and heating. A low level of diastase activity could indicate that the honey is old or that it has been overheated, or both.

The other component is *hydroxymethylfurfuraldehyde* (HMF) which is produced by the chemical breakdown of fructose, it is measured in milligrams of HMF per kilogram of honey. It is quite harmless in the amounts present in honey. Production of HMF is a continuous one and starts as soon as honey is produced, so the amount present increases with the age of the honey; it also increases with an increase in temperature (see graph). Thus honey which has been overheated during the handling process will show a low level of diastase activity and a high level of HMF. These changes can occur if honey in metal drums, especially if of a dark colour, are exposed to strong sunlight.

Bees also collect honeydew and process it like nectar, into honey. There is a lot of misconception about the production of honeydew which is a secretion from the bodies of certain homopterous insects such as aphids and coccids (scale insects). These feed by having tough sucking mouthparts, probosci, with which they pierce the sieve tubes of phloem tissue of plants. The pressure of phloem sap is so great that it is forced up the proboscis and into the atimentary canal of the insect,

13

H.M.F. content per kg honey per day

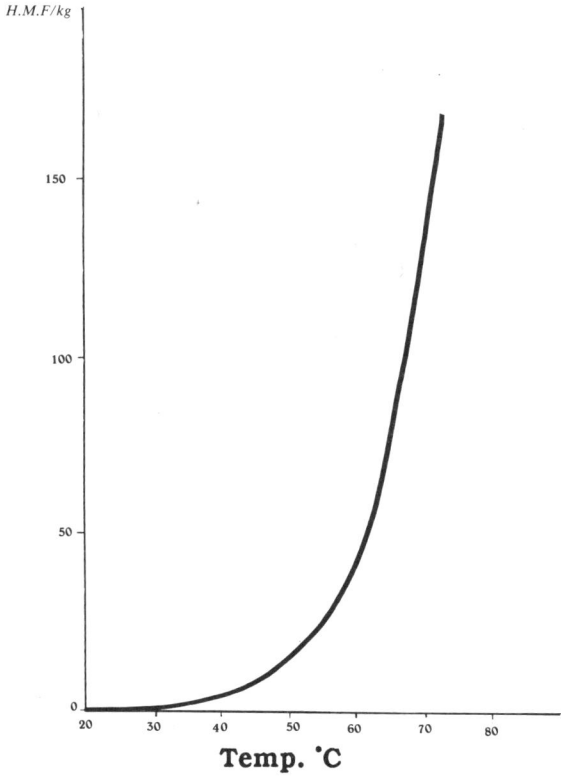

Fig. 4

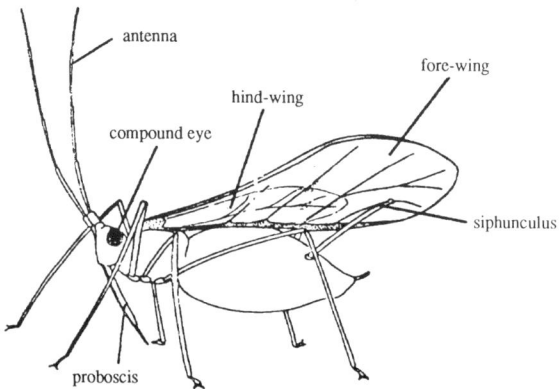

Fig. 5 Diagram of an aphis showing siphunculi

14

often at a rate far greater than it can cope with. The excess passes out of the alimentary canal via a pair of tubes the *syphunculi* or *cornicles* whose ends are situated on the dorsal side of the abdomen towards the tip. This secretion does not emerge from the anus like faeces. It has almost the same composition as phloem sap from which nectar is processed. Honeydew collects at the ends of the *syphunculi* in droplets and may fall off and coat the leaves of the plants on which aphids or coccids are feeding; it may fall to the ground. Honeydew honey differs little from floral honey except that the level of undertermined material may be as high as 10%. Such honey produced from coccids feeding on conifers is highly prized in Southern Germany where it commands a much higher price than floral honey.

As stated earlier at the beginning of this chapter, bees will collect any sugary substance, take it back to the hive and process it. Not only do they collect nectar and honeydew but anything that is sweet. Although they will process, store and seal it, much of these substances bear little resemblance to honey because they lack those materials which contribute so much to the delicate flavours and aromas of honey because they are not of floral origin.

These substances are mentioned because beekeepers and judges might be at a loss to account for green or pink honey or one having the flavour of Coca Cola. The history of green honey which appeared in Yorkshire and Norfok is of interest if only to tell the circumstances which led to it. In both cases in the making of chocolate filling at sweet factories in York and Norwich the mix which had been coloured green was sub-standard and had to be rejected. It was sold as an additive to cattle food to local farmers. Either because it was spilled in transit or stored where bees had access to it, it was collected, processed and stored, much to the amazement of the beekeepers who extracted the supers later. Pink honey, in Essex a few years ago, resulted from bees foraging in emptied marsh-mallow containers, while bees alarmed sightseers and holidaymakers in a London park by collecting the remains of Coca Cola left in discarded cans. In the period immediately after World War II the then Ministry of Food stored large quantities of West Indian sugar in hessian sacks on the runways of disused airfields. Although those sacks were covered with tarpaulins they were not entirely protected from the weather and the sugary syrup which oozed from them was avidly collected by foraging bees. Sometimes during judging I can detect the flavour of molasses. Is there a connection that farmers use molasses in the treatment of grass to convert it into silage?

Chapter 3:
Extracting and Handling Honey

Many years ago my mentor and long-time friend, Fred Richards, who at that time was CBI for Norfolk, said in the course of a series of lectures that there was nothing a beekeeper could do to improve honey after it had been extracted, but much he could do to spoil it. With one possible exception those remarks are true today for there is nothing purer and more wholesome than sealed honey on the comb. From the moment the extraction process starts honey becomes contaminated with undesirable substances ranging from pieces of wax, parts of bees, dirt, fibres from straining cloths and air bubbles, the latter incorporated into the honey as it is flung out of the combs during extraction. All these substances have to be removed if the honey is to be of saleable quality or presented for judging on a show bench.

Before honey is removed from the hive it should be sealed, this indicates to the beekeeper that the combs contain honey not nectar. At the end of a nectar flow or the end of the season some cells may not be filled and consequently not sealed. Usually it is perfectly safe to extract such honey. Because of the tendency of oilseed rape honey to granulate on the comb it is almost imperative that honey from this source is removed as soon as possible. Unfortunately some rape honey is often removed before it is fully ripe and fermentation will surely result if such honey is removed and bottled. There is a simple way of testing the contents of the cells and that is to hold the frames firmly in both hands in a horizontal position and give them a vigorous shake; any nectar or unripe honey will immediately fall out. These frames should be put into supers separate from those containing ripe honey.

Fig. 6 A frame of sealed honey

It is quite possible to remove excess water from unripe honey if the supers are put into a room in which the temperature can be raised to about 95°F/30°C and air can be circulated over the frames. This means placing the super in piles on slats and staggering them. If a fan heater is used and there is an extractor fan which can be used periodically to remove the moisture laden air, up to 1% moisture can be removed in 24 hours. You can overdo this treatment by prolonging it to the point where the honey becomes so vicious that it will not extract. Care must be taken not to raise the temperature too high or the combs in the top supers will collapse under the weight of honey as the wax softens. It is good practice to reverse the supers twice a day. Many beekeepers use heated rooms as part of the normal pre-extraction process as warm honey is easier to extract and less is left in the combs than if the honey is cool or cold.

Not all beekeepers have purpose built rooms in which to do the extracting and may have to use part of the house, usually the kitchen as there are work surfaces and a ready supply of hot and cold water together with a sink. Wherever the extracting is to take place ensure that it is beeproof. Occasionally beekeepers complain that although they enjoy working with their bees they do not enjoy the hassel of removing the crop. This is probably a reflection of the conditions in which they have to work. With a little forethought and good planning extracting honey in the kitchen need be neither a daunting nor messy task.

Fig. 7 View of honey house

Irrespective of the size of the task or the accommodation, honey extraction requires careful thought and it is a useful exercise to take stock of what has to be done, what appliances are to be used and the order in which the various operations take place, otherwise there will be frayed tempers a long time before the task is completed. It will be impossible to avoid spilling some honey and wax so ensure that all surfaces can either be cleaned and, if necessary, scraped easily or are covered with old sheets or newspaper. This can be a tedious business if extraction is done over a long period so it is easier to have duckboards on the floor and slats on the work surfaces where supers will be stood. The floor and work surfaces need only be cleaned when all extraction has finished.

If adapting an existing building or constructing a honey house ensure that it is beeproof, has adequate lighting, both natural and artificial, enough power points correctly sited, sufficient work surface and, if possible, an extractor fan.

When considering uncapping, extracting and straining equipment the choice may have to be made on the funds available so only guidlines are given. Uncapping combs can be a messy business so can the subsequent treatment of cappings and uncapped combs awaiting extraction.

Uncapping devices range from plain and serrated edged knives through steam and electrically heated knives, uncapping planes and purpose made uncapping machines as used by large commercial enterprises. The average beekeeper, even those having 100+ hives can cope quite well with the most inexpensive uncapping equipment - the serrated edged kitchen knife, used unheated; it is a fallacy that heated knives are needed to uncap honey.

On a small scale cappings can be dealt with by having a piece of wood straddling the top of a clean washing-up bowl. A 2" nail hammered through the centre of the piece of wood will act as a secure anchorage for the lug of the frame being uncapped. The cappings can subsequently be strained through whatever material is being used for straining honey, while the strained cappings can be rinsed in cold water and subsequently rendered into cakes of wax while the rinsings, if sufficiently concentrated can form the basis of mead. While waiting extraction the frames can be stood top bar downwards on a suitable sized rectangular tea tray.

Probably the easiest way of dealing with cappings is by using a 'Pratley' type uncapping tray, an example of which is shown on the next page.

Basically this type of tray is a container of water heated by a 1000 watt kettle element and having a sloping surface. Filling with water and the release of any steam produced is through a slot shaped aperture at the end where the slope starts, while at the other end of

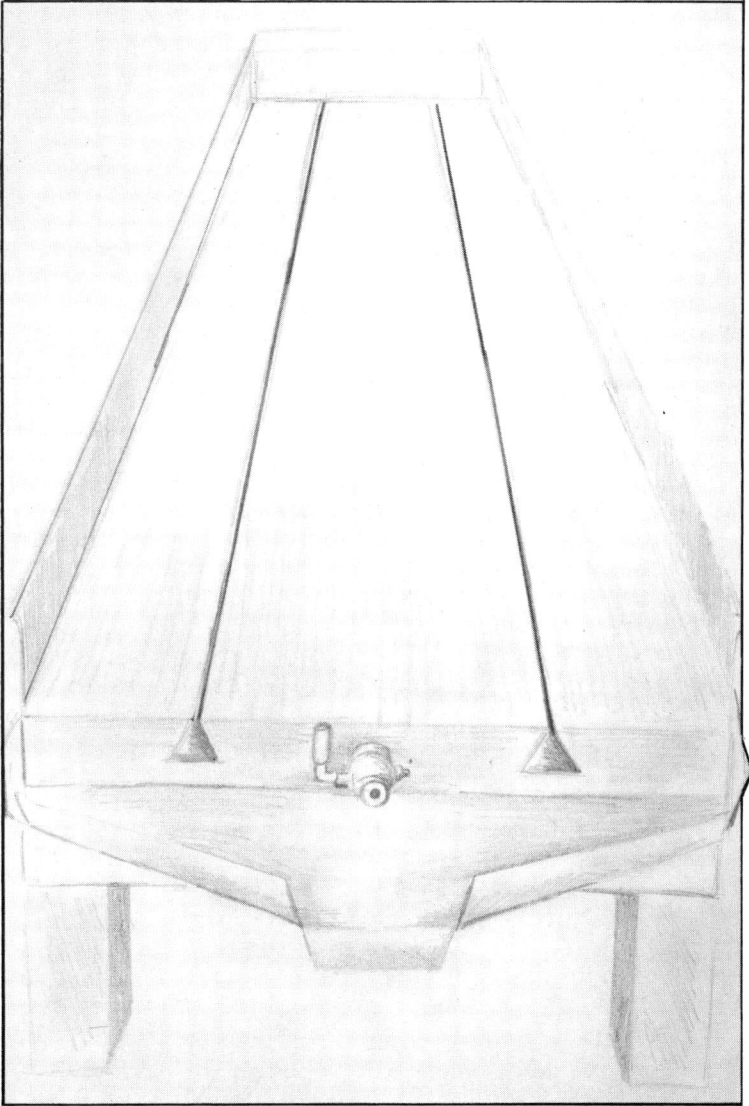

Fig. 8 Pratley uncapping tray

the slope is a spout through which honey and melted wax can drop into a suitable container. A strip of wood with one, sometimes two recesses in the top side straddles the top of the tray; the recesses take the ends of lugs when uncapping is being done. A perforated strip above the spout prevents any solid matter leaving via the spout. It is sometimes suggested that the honey from the cappings is over-heated, although it is usually on the surface of the tray for such a short time that little if any damage is done. Because the honey becomes mixed with pollen on the surface of the tray, honey from the tray is usually cloudy and unless filter pressed is not presentable as liquid honey. The best way to use such honey is to put it in separate labelled containers and cream it. Incidentally, as the honey and melted wax fall into the container, the wax being lighter rises to the top and as it cools it forms a solid cake which can later be lifted off leaving the honey. Up to six frames can be stood, top bars downwards, on the uncapping tray if the wooden strip is positioned near to the spout end. Any honey which falls out drops on the tray.

There are other devices for dealing with cappings and these are usually stocked by dealers in beekeeping appliances, they vary from uncapping trays and baskets usually made of plastic to larger un-capping tanks where the cappings drop on to steam heated pipes. Some beekeepers place the wet cappings in Miller type feeder on hives. Whatever method is used for dealing with the cappings remember that beeswax is a very valuable by-product of the extract-ing process and under no circumstances should it be wasted. There are better returns on beeswax than trading it in for equipment.

Extractors vary from the small table top variety taking two combs to the very large type used by commercial beekeepers taking up to 50 frames or more. They are of two designs, those which extract with the frames placed tangentially or those having the frames arranged like the spokes of a wheel - radial extractors. The power to drive the cage could be manual or mechanised, in other words hand or power driven, usually by an electric motor; if the latter it is usual to have a speed control. The materials used are usually polythene or stainless steel; older extractors were made of tin plate or monel metal, if the former great care has to be taken to ensure that all traces of honey are removed; the extractor is dried and given a thin coating of liquid paraffin prior to storage over the winter, to prevent rusting.

All extractors use centrifugal force to throw the honey out of the cells. The advantages of tangential extractors are that they are usually cheaper to buy and the frames do not need to be rotated so fast; they are usually smaller than radial extractors. On the other hand they only extract up to six frames at a time and the extractor has to be stopped and the frames reversed as only one face of the comb is extracted at a time. Usually the frames have to be changed round by hand but some machines do this mechanically by placing the frames in baskets which are pivotted. Because of the forces acting on the combs it is advisable to partly extract one side, reverse the frames and fully extract the other side, reversing again to completely extract the first side. This process is tedious and time consuming especially if newly built combs or older, well filled and warm combs are not to be damaged. Probably little is gained by placing the frames so that the bottom bar leads when rotated. The theory behind such positioning is that more honey is extracted because the cells slope downwards.

Fig. 9 Tinplated tangential extractor

The higher speeds needed to extract the honey with radial extractors means that gearing has to be used in hand operated machines. Since the frames are arranged like spokes of a wheel and larger numbers can be extracted, radial extractors are larger than tangential ones. These are the main reasons for such extractors being dearer.

Extraction using radial extractors is a slower process as there has to be a slow build up of speed to about 250r.p.m., sometimes more. The frames must be placed with the top bars away from the centre. The obvious advantages of radial extractors are that more combs can be extracted at a time and there is no need to reverse the frames.

Some manufacturers stock screens which can be placed in radial extractors to convert them to tangential – useful when extracting from newly built combs – but they add to the cost. At least the user has the best of both worlds.

If you have only a few combs to extract and don't mind sacrificing them, they can be cut from the frames and the honey squeezed out through suitable straining material or pressed out using hand squeezers stocked by some suppliers.

As previously mentioned, oilseed rape honey presents problems if it is not extracted as soon as possible. Probably no beekeeper whose bees have worked this crop has not had some granulated honey to deal with. One thing should not be done to it and that is to leave supers of this honey on hives as the main source of winter feed. Bees cannot use such honey unless it can be liquified and that means the bees have to forage for water – and in mid winter. Even if they could do this and that is only possible on odd occasions, the amount they could bring back would be insufficient for their needs; they would die in the midst of plenty.

Some beekeepers have tried to liquify this honey by using heat, all attempts seem to have been disastrous as the wax has softened and the combs have collapsed under the weight of the honey in them. If small amounts only have to be dealt with a kitchen tablespoon can be used to scrape the combs down to the mid-rib, it helps if the edges of the spoon have been filed sharp. The scrapings can either be dropped onto a Pratley tray or put in a container to be heated later. If the whole comb or the greater part of it is granulated hard it is best to remove the comb from the frame and treat it as scrapings are treated. To separate the wax from the honey it is necessary to first melt the former, this has to be done by heating the containers to about 149°F/65°C temperature just above the melting point of beeswax. This is not a difficult process if a thermostatically controlled warming cabinet is used. The contents of the containers should be stirred periodically – every six to twelve hours, to ensure that no hot spots develop and the honey becomes overheated. When all the wax has melted the containers can be left to cool when the cake of wax on the top can be removed and the honey strained. In all cases where honey has granulated in the comb, the comb has to be sacrificed or damaged in order to remove the honey. Comb which has been scraped to the mid-rib and probably damaged in the process is readily repaired and rebuilt the following year; use narrow spacing or alternate the frames with those having drawn out comb.

Heather honey, being thixotropic is usually extracted by pressing it out of the combs; as with oilseed rape honey which has granulated, the combs can either be scraped down to the mid-rib or removed from their frames and placed in suitable straining material to withstand the pressure required to press the honey out of the combs and through the strainer. Small cider or wine presses are quite suitable for this purpose but most appliance manufacturers offer honey presses of one type or another. Small domestic spin dryers will extract heather honey very well, all that is required is a cage made of dowelling or similar material inside the drum. Heather honey extracted this way normally contains large amounts of very small air bubbles.

Fig. 10 Top view of a radial extractor showing placing of frames.

Fig. 11 Small scale honey press.

Heather honey can be extracted in radial or tangential extractors if the cells are first uncapped and the honey agitated to make it somewhat runny, this method avoids destruction of the combs. Heather honey rollers or devices called perforextractors can be used to do this task manually while for the larger operator the task can be done mechanically by machines produced in Scandinavia.

For the small beekeeper heather honey is probably best dealt with by bottling it up as soon as extracted, but large amounts can be put into containers and bottled up as convenient. Pure heather honey granulates very slowly and with large globular granules, but if it contains even small quantities of other honeys it granulates fairly quickly and may need care when heating it to remove the granulation.

Small amounts of floral or honeydew honey are often bottled up soon after extraction, having been strained and allowed to stand for the air bubbles to rise. There can be problems with this procedure as cold honey does not strain readily and if it contains even small

24

amounts of honey which granulates readily, it may go solid in the honey tanks or even in the extractor if left in overnight. It is far better to run the honey from the extractor into plastic buckets or honey tins and warm it to remove any incipient granulation and reduce the viscosity. Not only does this ensure easy straining but air bubbles will rise to the top much quicker. The other problem which arises from bottling the honey almost directly after extraction is that the producer has little or no control over the final product, which is almost certain to granulate, often with an unattractive granulation. The market demands liquid and granulated honey at any time during the year and the beekeeper ought to equip himself to meet such a market; in addition to that, consumers are not likely to return to purchase granulated honey which is so hard that it is difficult to remove from the jar, impossible to spread and which probably has sugar crystals resembling grains of sand in size and texture.

For the owner of ten or more hives all extracted honey should be run into airtight containers. They may be lacquered honey tins or plastic buckets with lids; the latter are to be preferred as almost all honey can be removed from them, an impossibility with tins, they are also easier to clean and do not rust. This problem of storage can be overcome by running the honey into plastic bags inside a suitable container. The bags can be made airtight with the usual tie or a rubber band. Since the honey does not touch the insides of the containers they do not need washing. Great care should be taken during storage to ensure that HMF values do not rise either through long storage or high temperatures.

Chapter 4:
Preparing Honey for Sale and Show Bench

In theory, and it is not so difficult in practice, there should be little difference between honey presented for sale and that exhibited on the show bench, except that honey for sale will probably be a blend of several extractions while the honey selected for the show bench will be that having the best density, aroma and flavour. For this reason sufficient honey from each extraction or source should be put on one side and the best selected a week or so before showing.

The problems of straining cold honey direct from the extractor were mentioned in the previous chapter and enlarged upon now. Two materials are most commonly used, muslin and nylon cloth of fine weave, usually called organza. Of the two, nylon is to be preferred as muslin being spun from cotton fibres often sheds these fibres into the honey when strained; its mesh can also be distorted much more easily allowing large particles through. Whichever material is used it should be fashioned into a sleeve so that it almost touches the bottom of the honey tank, by so doing it presents the largest surface area, has the least strain put upon it and is less likely to add to the air bubbles that can be incorporated into the honey at this stage. A method of fastening must be used so that the strainer does not come adrift; tying the strainer with string is rarely adequate, two methods which usually suceed are either by securing it with steel wire tightened with a pair of pliers or a wooden or plastic ring of a suitable size as used for embroidery. These can be purchased from needlecraft or wool shops. If handling large quantities of honey a straining tank can be set aside for this purpose only and the strainer can be used without changing for a whole season.

For the large producer there are honey-strainers which use a series of cylindrical screens of decreasing mesh size through which the honey passes. The best known type is the O.A.C. strainer developed by the Ontario Agricultural College in Canada. This allows continuous bottling to take place.

The easiest way of straining honey in quantity and subsequently marketing or showing it as liquid or granulated (creamed) honey is by using a warming cabinet or a warming room in which the honey can be placed and heated to the desired temperature. A number of publications including "Honey from Hive to Market" now long out of print, give details of honey warming boxes, usually these were designed around 28lb honey tins. To accommodate 30lb. honey buckets they need to be slightly larger.

Fig. 12 Lacquered honey buckets

Any suitably insulated container, even redundant domestic refrigerators will do as the basis for a warming cabinet. All that is required is a suitable method of heating and a means of controlling the temperature. Electric lamps are quite adequate to heat a cabinet taking up to 4 x 30lb. honey buckets and the temperature can be controlled by using lamps of the appropriate wattage. Better control of temperature and a more rapid heating can be obtained by incorporating a rod thermostat in the heating circuit especially as different temperatures are needed for different processes.

Warming cabinets are available from some appliance manufacturers but there are usually small and inadequate for dealing with large quantities of honey. The average handyman can construct one based on plywood and expanded polystyrene. It is an advantage to have top loading as hot air will not escape so readily when the lid is removed, as happens with a modified domestic refrigerator. Although the plan shows expanded polystyrene as an insulator any material with similar properties, newspaper, shavings, sawdust, etc. can be used.

By using temperatures in the range 110°F to 120°F/43°C to 49°C not only can honey be warmed prior to straining but granulated honey can be liquified and liquid honey be prevented from granulating for several months. Commercial beekeepers and honey packers

Fig. 13 Honey tank strainer

Fig. 14 Top view of honey warming cabinet.

Fig. 15 Details of interior of warming cabinet.

often use higher temperatures to speed up operations. Mixtures of wax and honey such as oilseed rape combs containing granulated honey require much higher temperatures as previously mentioned to melt and separate the wax.

Care must always be taken to ensure that the honey is not subjected to long periods of high temperatures because of the effect on diastase activity and HMF content. The same changes in levels of the latter which take 300 days at 68°F/20C take 60 days at 90°F/32°C, 3 days at 125°F/52°C and only 4.5 hours at 160°F/71°C. Thus the necessity for avoiding high temperatures and for stirring granulated honey when liquifying it. (See graph on page 14.)

To the average purchaser of honey there are three sorts, runny, solid and comb honey. To the producer liquid, naturally granulated or creamed, and section or cut-comb honey. Occasionally chunk honey is offered for sale or is seen on the show bench. In recent years sections of honey have become a rarity and are often only seen on the show bench. Because special management techniques are needed to produce good comb honey a separate chapter is devoted to this.

Liquid honey after it has been strained from the warming cabinet should be transferred to honey tanks and allowed to stand for 48 hours before bottling, this ensures that the majority of air bubbles have risen to the top. After bottling, such honey should be returned to the warming cabinet for 12 to 24 hours at a temperature not exceeding 95°C/35°C, this allows the tiniest air bubbles to rise. For sale to the public no further treatment is needed but for show purposes any remaining bubbles should be removed from the surface of the honey and under the shoulders of the jar with a teaspoon. The inside of the jar above the honey should also be cleaned with a piece of cloth wrapped around a matchstick. Packers of large quanities of honey filter it using distomaceous earth which removes all solids including pollen together with air bubbles. Such honey often loses a little of its flavour as it is heated to 140°F to 160°F/60°C to 71°C prior to filtering, after which it is cooled rapidly. Providing the water content of liquid honey is below about 20% it will store indefinitely at room temperature 60°F/15.5°C.

In the British Isles there is a greater demand for granulated or creamed honey than there is for liquid honey. Because of the problems associated with naturally granulated honey it is preferable to produce creamed honey for the market as the beekeeper/packer has a large measure of control over the finished product. All that is required in addition to the method used for producing liquid honey is a supply of suitable honey for seeding and a means of incorporating it thoroughly into the liquid honey to be creamed. Oilseed rape is ideal for seeding because of its very fine granulation.

Having strained the liquid honey it should be stood in suitable

Fig. 16 Plans for honey warming cabinet

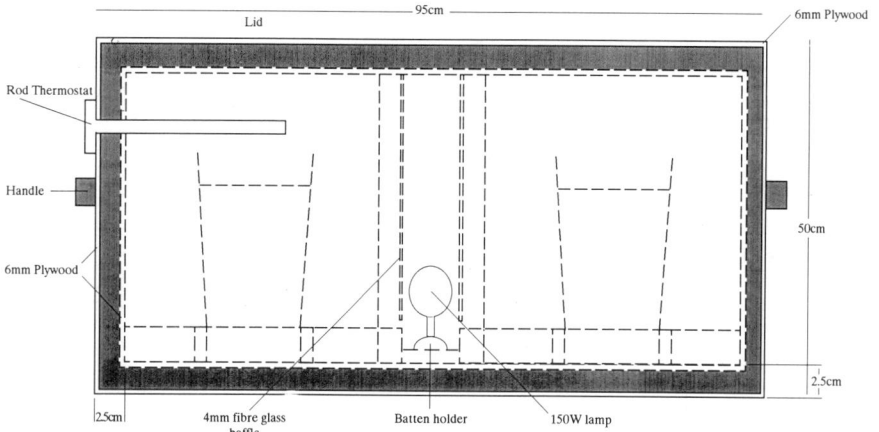

Side Elevation

The basic structure is 2.5cm (1") square timber, butt jointed and covererd with 6mm plywood. This gives the warming cabinet strength and lightness a need for complicated joints. The cavity between the inner and outer plywood skin is insulated, preferably with expanded polystyrene.

Warming cabinet for 4 x 30lb buckets

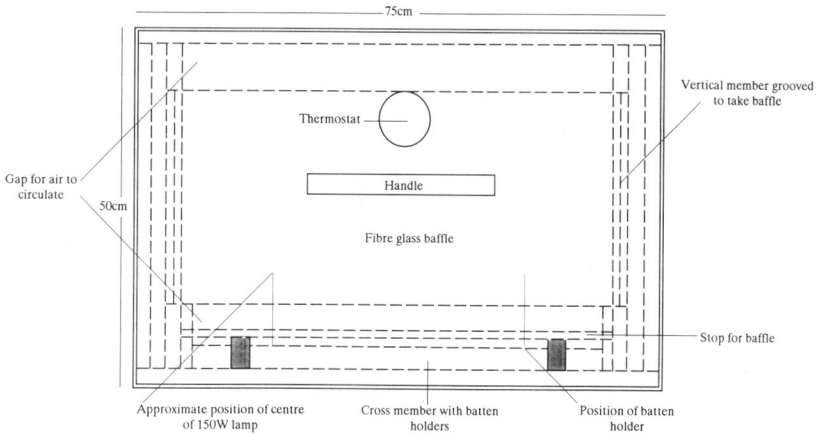

End Elevation

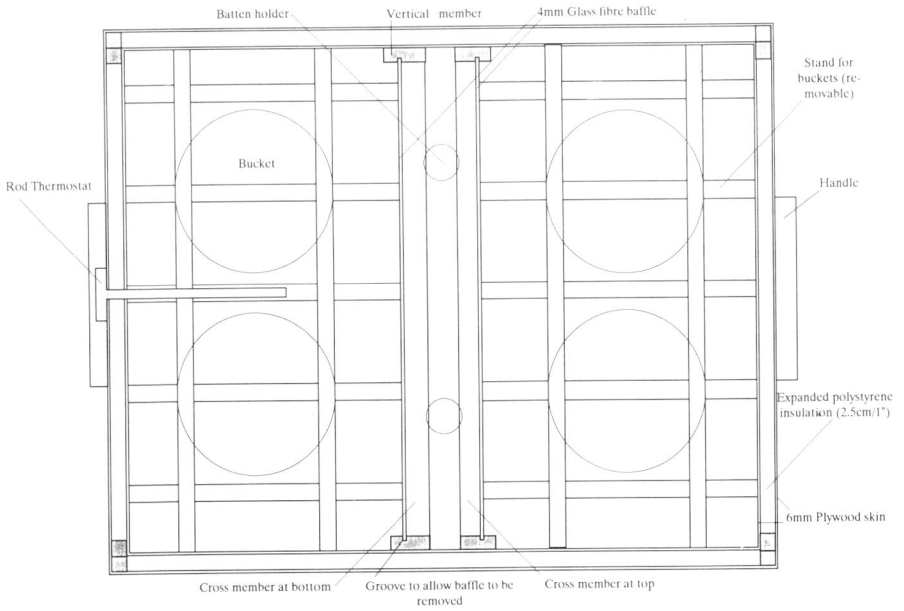

Batten holder   Vertical member   4mm Glass fibre baffle

Stand for buckets (re-movable)

Rod Thermostat

Bucket

Handle

Expanded polystyrene insulation (2.5cm/1")

6mm Plywood skin

Cross member at bottom   Groove to allow baffle to be removed   Cross member at top

Plan view with lids removed to show internal construction

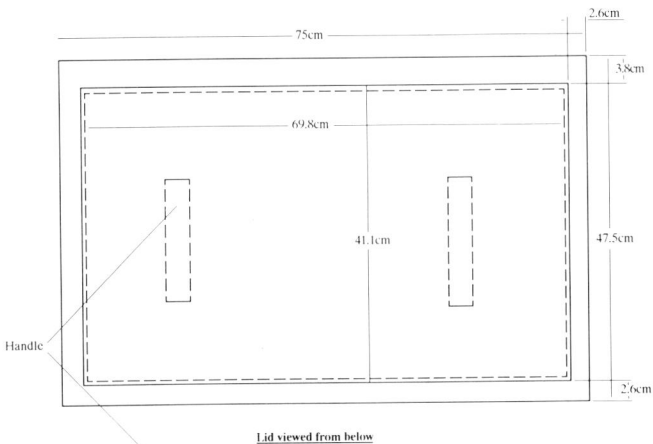

75cm

2.6cm

3.8cm

69.8cm

41.1cm

47.5cm

2.6cm

Handle

Lid viewed from below

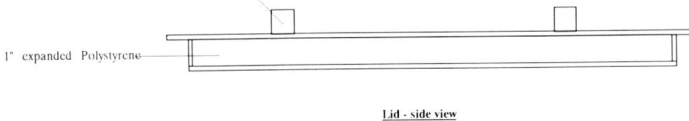

1" expanded Polystyrene

Lid - side view

containers for creaming. Suitable containers can be honey buckets or honey tanks depending on what process is used to incorporate the "seed" honey. The method used for creaming honey is known as the Dyce process after Elton James Dyce who, when as professor of apiculture at Ontario Agricultural College (now Guelph University), developed and patented the process in 1931. By this process the honey is allowed to cool to about 75°F/24°C. Higher temperatures could cause the crystals of the seeding honey to dissolve, while lower temperatures would increase the viscosity of the honey and make thorough mixing very difficult. Usually 10% of the honey previously selected for creaming is added and thoroughly mixed. This can be done on a small scale by using a stirrer such as sold by appliance manufacturers, or a paint mixer rotated by an electric drill. Great care must be taken to ensure that air is not incorporated into the honey as it may subsequently cause frosting. On a very large scale seeding and liquid honeys in separate containers are mixed in the correct proportions as they pass to a third tank from which the honey will be bottled. After creaming the honey is left to stand in tanks for about 24 hours before bottling.

Fig. 17 Equipment for creaming honey

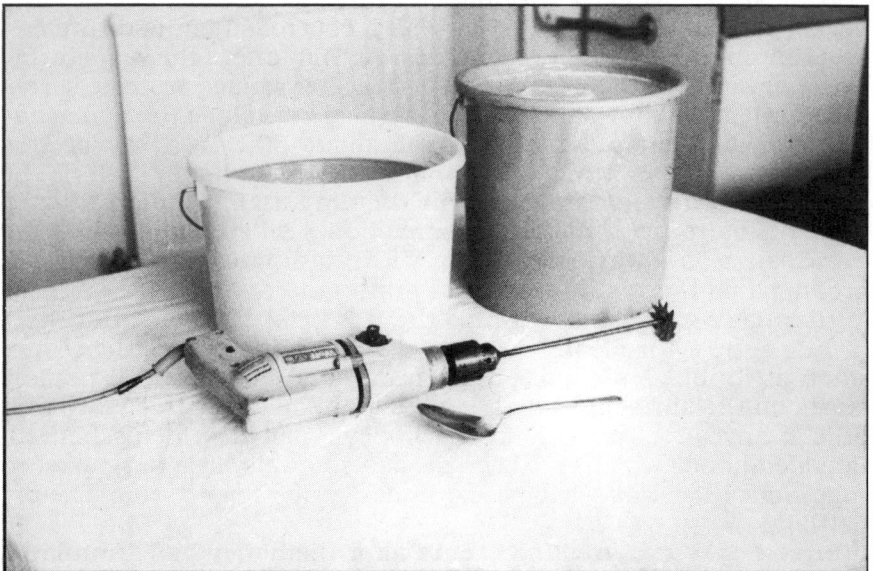

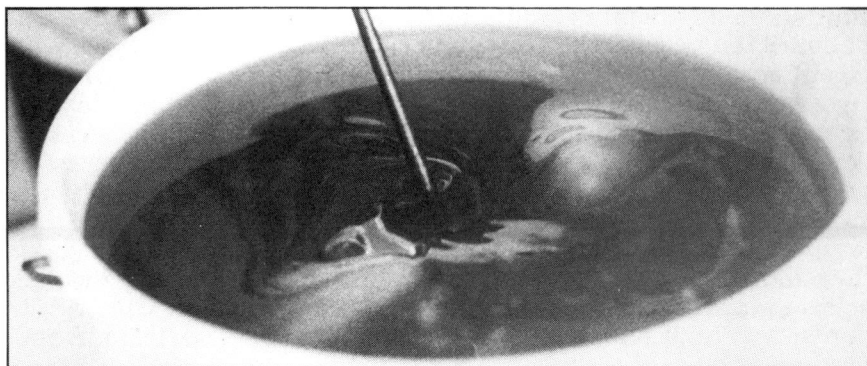
Fig. 18 Mixing under way

Creamed honey can also be produced by warming granulated honey at about 90°F/32°C, stirring it thoroughly at intervals to ensure the honey is thoroughly mixed, then bottling it.

By whichever method honey is creamed it should be kept at a temperature of 57°F/14°C for about three days after bottling to allow rapid crystallisation of the glucose. Temperatures below 50°F/10°C or above 50°F/15°C will result in the honey taking longer to crystallise. Obviously this temperature may be difficult for the beekeeper who has no store in which a controlled temperature can be maintained, especially in summer. At that time of the year storing the honey outside either in a shed or in a shady place protected from the weather is about the best that can be done. Once the honey has crystallised it should be stored at about 50°F/10°C as at this tempeature yeasts, which occur naturally in the honey, will not be able to cause fermentation, bearing in mind that granulated honey ferments more readily than the same honey when liquid.

Frosting is always a problem with granulated honey whether produced as the result of natural granulation or creaming. Some authorities recommend warming the jars before bottling, but there is little or no evidence to show that this is effective. Frosting occurs when air bubbles are present in granulated honey and it usually starts on the surface, so great care should be taken to ensure that little if any air is present in the honey when it is being bottled. Provided honey which has been seeded contains little or no oilseed rape honey it can be left in the honey tanks for 48 hours before bottling.

Another type of frosting occurs after the honey has granulated and it is subjected to low temperatures which cause the honey to contact away from the glass leaving an air gap which rapidly frosts within a few days. This is one of the reasons why granulated honey is often packed in opaque plastic or waxed cardboard containers. The customer cannot see the frosting, which incidentally does not affect the flavour of the honey.

Those beekeepers and packers who have the facilities to store granulated honey at a constant temperature are less likely to have the same problems with frosting that the smaller producer has, so it is unwise to bottle larger amounts of creamed or granulated honey than can be sold in about a month. For the same reason the exhibitor is advised to cream honey no more than a fortnight before it goes on the showbench. Jars of naturally granulated or creamed honey need little attention before showing except, as with liquid honey, the inside of the jar above the level of the honey should be cleaned.

Apart from keeping examples of each extraction and/or source of honey separate so that the best can be selected for showing, and applying the finishing touches to the exhibits, honey for showing should differ little from that offered for sale. The only other things that exhibitors have to do is to read the show schedule very carefully so that all entries are made in the correct classes. A final polish of the jars and a last minute change of lids to ensure nothing offends the judge is all that can be done. The rest is up to the honey itself.

Chapter 5:
Comb Honey

To a very large extent the quality of finished comb honey, whether for sale or show-bench and in whatever form, depends on the management techniques used. It is also dependent on the weather conditions and the availability of nectar as the foundation has to be drawn, filled and sealed in the shortest possible time. Foundation will not be drawn out by any but strong colonies, nor in the absence of a good nectar flow and very favourable weather conditions. The quality is also affected by the nature of the honey, which should be liquid and disinclined to granulate. When comb honey has granulated not only does it look unattractive but to most would be purchasers it is unacceptable; to show judges this is most certainly the case. With the increase in the acreage of oilseed rape in recent years the production of good comb honey has become very difficult in those areas where this crop is grown.

There is nothing more pleasant to the eye, more delightful to the palate nor more natural than honey on the comb. Because it comes direct from producer to consumer looking for a 'natural food'. As such it commands a higher price than the equivalent weight of extracted honey.

The author's interest in beekeeping stems back to when he was about five years old, when having walked the two-and-a-half miles back home from school on a warm summers day he was plied with honey from a section costing 6 old pence, 2.5 pence in decimal coinage, liberally spread on bread and butter. Those were the days in the mid 1920's when white clover abounded in the meadows and charlock in the corn fields. Skep beekeeping was still widely practiced by villagers and small farmers to augment their very low incomes. How things have changed!!

Section Honey

For economical reasons very little comb honey is produced in sections today in the British Isles. The wooden square or plastic round sections require very strong colonies, ideal weather conditions and very heavy nectar flows before the bees will work them. Very strong colonies should have all supers removed and a section crate put on at the beginning of a flow, the crowding which results will force the workers to draw out the foundation and store in the sections; such crowding usually results in swarming if regular inspections and swarm prevention methods are not used. Very large swarms will also work sections for a short period after hiving as there is no brood present, but as brood appears and the number of foraging bees decrease through normal mortality the ability of swarms to produce good section honey is of limited duration even when fed heavily to encourage wax production.

Fig. 19 Section Honey

Sections can be placed on or in the hives either by using section crates or in hanging frames in the supers of extracting honey. The British National section crate holds 32 sections, eight rows of four in each row. Each row of sections is separated from the next by a slotted metal divider, the slots so placed to allow the bees access to each individual section, this ensures that the cells are drawn out a certain distance and the cappings are flat and regular. Hanging frames hold three sections each and as with a section crate the sections are separated from the surrounding combs by metal dividers. Bees can often be encouraged to occupy section crates if one or two partly drawn out sections are placed in the middle of the crate. It is unwise to leave sections in the hives after a flow has finished as both wood and wax become travel stained and unattractive.

To obtain a maximum crop from heather it is necessary to use strong colonies headed by young queens. These conditions, together with the fact that such colonies are unlikely to swarm and have brood chambers full of brood and eggs, provide the best chance of obtaining good sections. There is also no danger of contamination with oilseed rape honey.

Sometimes the cappings of comb honey are disfigured by the tunnels of the larvae of the Bee Louse (*Braula coeca*), not actually a louse but a wingless fly. With severe infestation of this insect, which appears to cause no harm to bees and might be mistaken for the parasite mite, Varroa Jacobsonii, the tunnels can be very obvious and make the entry unacceptable to a judge. Some judges appear to accept a small amount of tunnelling without any penalty ensuing.

Fig. 20 Comb disfigured by Braula

When selecting and preparing sections for sale or exhibition ensure that, as far as possible, they are well filled and fully sealed. Any propolis on the woodwork should be removed. Most sections are presented for sale in cardboard cartons having a transparent cellophane window on one side. The container must carry the name and address of the producer or packer and must show the nett weight of the section in imperial and metric units. Some sections are sold in transparent cellophane wrapping, but this is unusual in the British Isles.

For show purposes it is essential to have all cells filled and the cappings complete and even. The judge usually gives preference to the lightest wax. All woodwork should be scrupulously clean and cells should contain no pollen or granulated honey, both can be detected when a strong light is shone through the comb. The judge will usually sample the contents of at least one cell – usually close to the edge. Sections are shown in glass-sided cases with "lace" edging.

Cut Comb Honey

The disadvantages of producing sections can be overcome by using shallow extracting frames in which have been inserted extra thin, unwired foundation. The bees will draw out the foundation in the normal way and when sealed the combs are cut from the frames and further cut into rectangles 85cm x 65cm. Any comb which is unfit for sale as comb honey can have the honey pressed out and bottled in the usual way. If the combs are filled on wide spacing, nine frames to a British National super, each pack of cut comb honey should average 0.5 lb/227g. net weight.

The main problem with obtaining cut comb honey of the required thickness is obtaining even combs in the absence of spaces as used for sections. If frames containing the foundation are drawn out on wide spacing (1.7/8"/47mm) much 'wild' comb will result. To avoid this the foundation should be 'started', or the initial drawing out done, on narrow spacing (1.7/16")/36mm using 11 or 12 frames in a British National super. The frames should then be respaced, just before the honey is sealed to 9 per super. This is no problem on a small scale when metal ends or their plastic equivalent are used. This type of spacing, like the use of Yorkshire spacers, is not so widely used nowadays as the spacers have to be removed prior to extraction. For simplicity and ease of operation castellated strips have found increasing favour. Since these are fixed spacers and cannot be moved the use of two separate methods of spacing for comb honey presents problems which could be solved by duplicating supers, the first to be used having narrow spacing while the combs are finished in the second having wide spacing.

This problem can be overcome by using a single super inverted for drawing out foundation and the right way up for completion. Unfortunately, the British National super with bottom bee space has only 3/4"/6mm recess so the lugs of frames which are 7/16"/11mm would protrude. This can be circumvented by one of two methods, either by fixing four strips of wood 3/10"/5mm thick on the underside of the super or when constructing supers take 3/16"/5mm off the depth of the shorter sides. Any of three methods of spacing to accommodate 12 frames in the inverted super can be used, castellated strip, Hoffman or Manley spacing. As soon as the combs have been drawn out to the desired thickness, the frames can be removed, the super inverted to its normal position and nine frames for completion put in. The remaining frames can be put into spare supers taken along for that purpose. This method works very well and is not only labour saving and economical of equipment but is the only efficient method when working for comb honey on heather over 200 miles from home as many East Anglian beekeepers do every year with bees on the North Yorkshire Moors. All that is required is one trip to rearrange the frames at the appropriate time.

Another method is to have the foundation drawn out before the bees are taken to the moors. If this can be done the combs can be extracted using great care in a tangential extractor run at slow speed. If the honey is unripe it should be fed back to the bees. The drawback with this method is that the wet combs might contain honey such as oilseed rape which could cause rapid granulation of the heather honey and thus defeat the object of the exercise, unless they were first dried. All this causes additional labour which might be considered unnecessary.

Fig. 21 Reversible super for cut comb production

When considering cut comb honey for exhibition, the best frames can be used for showing as frames of honey suitable for extraction. The main points with these are that the woodwork should be clean, the comb completely filling the frame and with a level surface. It should be of sufficient thickness that the wax cappings can be removed with one cut of the uncapping device and then there should be no hollows. The honey should be liquid with neither pollen or granulation. Preference is usually given to cappings of light colour with few if any Braula tunnels. As with section and cut comb honey the judge will taste the contents of at least one cell. Such frames should be exhibited in cases having glass in each side. They should be easily removed from the case whose top should not be secured with nails as the judge will need to remove the frame from detailed examination. Whatever method is used to secure the top, be it screws or some other device, the top should be capable of being removed without testing the ingenuity or patience of the judge. The case is not usually judged but should not be of such poor quality as to influence the judge in his final decision. Many judges accept heather honey as being suitable for extraction.

For show purposes cut comb honey should be cut to the same dimensions as for sale and in all other respects should conform to the other requirements for good comb honey. The exhibits should be presented in the same containers as used for sale, white plastic boxes with airtight transparent plastic lids. With all forms of comb honey great care should be taken to ensure that the cappings are not punctured. If this happens atmospheric moisture will be taken up by the honey which will "weep" detracting from the quality of the exhibit.

Show cases for frames of honey, like those for sections are obtainable from many suppliers.

Chunk Honey

This type of honey is sometimes sold in shops and there are classes for it in some honey shows. Chunk honey consists of a piece of comb honey in a jar of liquid honey. The piece of comb honey should be large enough to reach from the bottom of the jar to the surface of the honey and just large enough in cross section to pass comfortably through the neck of the jar. Only comb honey which is unlikely to granulate and liquid honey having the same property should be used. For show purposes both the quality of the comb honey and the liquid honey are considered by the judge. For obvious reasons once the chunk honey has granulated it has lost its appeal, also it cannot be reliquified without damage to the comb it contains.

Chapter 6:
Marketing

Sticky jars, crocked and ugly labels, fermented or crystallised honey are all indecent and this exposure should be avoided at all costs. Underexposure can be disastrous if selling your crop is necessary. One of the best ways not to sell honey is not to let anybody know it is available.

Marketing is exposing a product for sale – keep your product and your exposure in top shape.

These are not the thoughts of the author but of a writer in a recent issue of "Gleanings in Bee Culture".

One might also add that selling honey at the garden gate at knock-down prices in a glut year when the honey is badly strained, packaged and labelled, and with a thick layer of air bubbles on the top makes the task of marketing home produced honey by reputable beekeepers and packers very difficult, especially since most imported honey is well prepared and presented.

At the present time after four indifferent summers and heavy losses of colonies during the years 1986-1988 inclusive, home produced honey has been scarce and there have been few if any problems selling it. It is a seller's market but such conditions have not always existed nor are they likely to continue indefinitely particularly if there is a glut in the world market and after 1992 when the E.E.C. will become an open market for its twelve member states.

Although at present indifferent honey indifferently presented may find a ready sale there is no excuse for presenting anything but a first class product attractively presented, for sale. In this respect the action of many beekeeping associations in staging commercial classes at honey shows is to be applauded; for much of what exhibitors learn in preparing such exhibits rubs off on the products sold to the public.

Marketing has to be considered from a number of aspects: the quality of the product, the cost of producing it, packaging, labelling and advertising. The quality and preparation of the product have already been dealt with. The only other point to make is the preference shown by the customers. It may be that some customers have a marked preference for a particular honey, be it oilseed rape or heather. If this is the case such honey must be extracted and bottled separately. Again more customers seem to prefer a granulated or creamed honey to liquid honey, there are also those customers who demand liquid honey when all the crop has granulated. Remember the customer is always right, also remember that if your product is labelled as coming from a particular source the labelling must conform to the various regulations governing the sale of products such as honey.

There are various methods of packaging honey, bottles – usually called jars, made of glass or plastic, plastic tubs and even tubes such as those used for toothpaste. By tradition honey sold in the British Isles is packaged in glass containers, there has been a resistance to transparent plastic containers. Plastic tubs and tubes seem to be used mainly in North America. The traditional method of packaging honey imposes considerable cost on the producer or packer and this must be passed on to the customer. There is a limited market for honey which has been put in airtight packaging and sold in rather expensive ceramic containers, often in the shape of a skep hive. These are usually well presented and one wonders if the purchaser is more interested in the packaging than its contents.

Although many beekeepers will sell their honey in jars containing 1lb/454g nett weight, there are those customers who, for one reason or another, prefer to buy in smaller quantities. Many senior citiziens find that 1/2lb/227g. is as much honey as they wish to buy at any one time. Much honey is also sold in 3/4lb/340g amounts. At agricultural shows and similar events there is a demand, especially from young people, for honey in 1oz/25g plastic containers. All these outlets are worth considering when marketing.

Although bulk sales, enough to satisfy one retail customer for six months or more, may seem very attractive at the time it is worth remembering that unless liquid honey has been heated to 160°F/71°C, granulation is likely to set in after a few months, this type of granulation is usually most unsightly with crystals of sugar forming at the bottom of the jar and slowly growing upwards. Granulated or creamed honey will frost particularly if subjected to sudden cold temperatures, as can happen in some shops over the weekends or in winter. Both faults render honey unattractive to customers and consequently sales will decrease. Such sales should thus be limited to the amount that can be sold in a month and arrangements made with the retailer to replace any honey which has lost its sales appeal.

Another aspect of producer/consumer relationships is that of ensuring a constant supply of honey throughout the year. This can be difficult if a run of normal years is followed by a run of poor years. It might be a wise policy to limit sales in the years of plenty and store the surplus for sale during the years of famine. Only with experience will the beekeeper be able to assess the available market and act accordingly. In the long run it all boils down to a good producer/customer relationship built on mutual trust.

The purchaser of any commodity if faced with two identical products at the same price will always purchase the one which is most attractive; this applies to honey as well as to anything else. If the two products are identical in all other respects it may very well be the more attractive label that influences the purchaser. Many beekeepers, usually because they are small producers with a limited market do not give much consideration to the label and often use the cheapest, which is probably the least attractive. Remember it is illegal to sell honey unless it is labelled – you can give it away unlabelled! Most suppliers stock a variety of labels; many of them are plain and unattractive despite their pretty colours, probably confusing as well becaues there seems to be no standard metric equivalent of 1lb, it varies from 453g, 453.6g, 454g to 455g.

For the enterprising beekeeper there must surely be something more inspiring than the standard county or country label. There is hardly a place in the country which does not possess an historically or geographically interesting location. Castles, cathedrals, ruins, water-falls, rock formation, the possibilities are legion. In the author's case, although the village of Walsingham means little or nothing to most people, about 250,000 visitors either as pilgrims or tourists come to see why it developed as the second most important place of pilgrimage in this country in the Middle Ages. It has regained that importance in recent years. Why should it be only the places that sell useless knick-knacks designed to empty visitors' pockets and purses that trade on the historical and religious importance of the place? All that was required was the use of a photocopier that enlarged or reduced, a photograph of the ruined east window of the Augustian friary which housed the shrine, a postcard with a bee on it with wings and legs outstretched, normally sent to farmers and fruitgrowers requesting their co-operation when spraying insecticides, a friendly art teacher who has a mastery of calligraphy and a commercial label firm eager for trade. Surely this is not beyond the ingenuity of most people.

Beekeeping associations could take a lead in this and produce an attractive label for use by members. Some associations, Norfolk is one, produce attractive labels quite cheaply which help to swell the associa-tion's funds.

Fig. 22 A personalised honey label

44

Fig. 23 Honey labels

There are many Acts and Regulations which govern the sale of honey in the British Isles and elsewhere in the world, including all countries in the E.E.C. Many of these Acts and Regulations govern the labelling of honey. They are too numerous and detailed for this publication and are listed in The Illustrated Encyclopadia of Beekeeping edited by Roger Morse and Ted Hooper. Only the most important are listed here, among them the Honey Regulations 1976 No. 1832. This brought the United Kingdom into line with the other countries in the E.E.C. It covers the composition and labelling of honey; Regulations 2 & 14 came into operation in December 1976 and the remainder in May 1977. The relevant E.E.C. act which was designed to harmonise the regulation of all member states was The European Communities Act 1972 under the Directive 74/409/EEC of 24 July 1974.

The Honey Regulations 1976 defines the various types of honey:-blossom, honeydew, comb and chunk honey, etc. It also states that a particular type of honey, e.g. clover or heather must be produced wholly or mainly from the specified species of plant/s. Also covered are the composition and preparation of honey, the restrictions on the use of the word honey and the labelling and description of honey including references to origin together with penalties and enforcements. Probably the most important part of these regulations is that which covers the manner of marketing or labelling, the relevant paragraphs (1) and (2) of the regulations are quoted as they are the most important for beekeepers selling honey in jars or other containers.

9 - (1)   Any statement required by regulation 7 (this refers to the labelling and description of honey) to appear on a label marked on, or securely attached to the container of any honey:-
(a) shall be clear, legible and indelible;
(b) shall be in a conspicuous position on the label marked on, or securely attached to, the container in such a manner that it will be readily discernible and easily read by any intending purchaer or consumer under normal conditions of purchase or use;
(c) shall not be interrupted by other written or pictorial matter where such an interruption might mislead the purchaser or consumer as to the nature of the honey;
(d) shall not be in any way hidden or obscured or reduced in conspicuousness by any other matter, whether pictorial or not, appearing on the label.
9 - (2)   The height of the letters in any statement referred to in paragraph (1) of this regulation shall be such as is not calculated by any undue or insufficient prominence to mislead as to the nature, substance or quality of the honey to which it relates.

Briefly these regulations make it imperative that the labelling must be clear and bold and of such a nature that it does not deceive.

Other regulations cover the quantities in which honey in prepacked form can be sold and the figures denoting the weight of the honey. Summarising all these regulations, honey prepacked for the retail trade must have the following information on the label:-

1.   The description of the contents, this can be Honey, Pure Honey, Pure Norfolk (or any other county or country) Honey, Clover Honey, Heather Honey, etc. Whatever its description it must be true and not designed to mislead.
2.   The name or trade name and address of the producer, seller or packer.
3.   The nett weight of the honey in both imperial and metric units.
The minimum size of the figures used to denote the weight depend on the nett weight of the honey as follows:-
2mm   for   1oz/28g.
3mm   for   2 or 4oz/57 or 113g.
4mm   for 8 or 12oz, 1, 1.5 or 2lb/227g, 340g, 454g, 680g or 907g.
6mm   for any quantity exceeding 2lb/907g.

There are two schedules attached to the Honey Regulation 1976, the first gives the method of determining diastase actively and the second lists the compositional requirements for both floral and honeydew honey, and cover reducing sugar (glucose and fructose), moisture,

46

sucrose, water, insoluble solids and ash content, together with acidity. Two points to note as far as honey produced in the United Kingdom is concerned is tha the water content of heather honey should not exceed 25% and other honeys 20%. The diastase activity should not be less than 8 and the HMF content not more than 40mg per kilogram of honey. Remember that the former is lowered and the latter raised by prolonged heating.

The Weights and Measure Act 1963 (Honey) Order 1977 which came into operation in May 1977 states that honey except for amounts of honey sold by nett weight of less than 0.5oz/14g, and excepting chunk and comb honey, can only be sold in quantities of 1oz/25g, 2lz/57g, 4oz/113g, 8o/227g, 12oz/340g, 1lb/454g, 1.5lb/680g or multiples of 1lb/454g. Comb or chunk honey can be sold in any nett weight providing all other conditions of labelling are adhered to.

At the time of writing it appears that by 1992 all weights will have to be given in metric units; it is not known if the nett weights given above will be allowed using the metric equivalent or if, as on the Continent, honey will have to be marketed in weights of 125g, 250g, 500g, 1kg or multiples of 1kg. If this is the case then larger jars than at present in use will have to be produced.

When looking at the regulations one is tempted to ask if the use of terms such as 'Organic' or 'Natural' are really true descriptions of the contents or are designed to mislead the purchaser or customer into believing that there is something special about the contents? The only thing special about many such honeys is that they are usually cheap, blended, imported honeys. Equally confusing and probably misleading are those labels which depict what is apparently an old English garden with a skep or double-walled beehive, while tucked away somewhere in very small print is the legend "A product of many countries". It probably says much for the contents that such ploys are used. Maybe the regulations governing the labelling of honey should be more strictly enforced or changed in such a way that such practices are illegal. Incidentally all honey is organic – it is a carbon compound, and the only natural honey is in comb form.

Advertising is a difficult subject to be subjective about, especially as good honey should sell itself. For the beginner the best advertisements are to ensure that the product is of high quality, well presented, and to start selling through local shops and through associations at honey and other shows. It does not take long for satisfied customers to ask for more. There is an added bonus from selling through associations as they normally retain only 10-15% for association funds when the usual discount to shops is 20%. Honey sold from home gives the best return as there is no discount, a suitable sign in the window or a well painted double walled hive with suitable words on it, placed in the front garden or outside the gate works wonders.

Large producers and packers require no advice on selling their products, but the small producer might find the "Yellow Pages" a useful way of advertising the fact that you are a beekeeper. It is very satisfying to receive letters from visitors, living elsewhere in the country, which state that a purchase was made at a certain place and would the seller be willing to supply similar honey by post? Unfortunately, the cost of posting and packing are prohibitive. Even more satisfying is the receipt of a letter from the other side of the Atlantic stating that the purchaser has never tasted such exquisite honey, far better than can be purchased in his own country. No, these are not fictional! It only goes to show that good honey, well presented, will sell itself.

Costing

It is of little use to produce a first class product which sells readily, if at the end of the day you make a loss. It is immoral to do so when other beekeepers, who have invested their hard earned money in small beekeeping enterprises, are trying to make ends meet.

The costing of honey is an exercise beset with many problems; for the commercial beekeeper or packer the problems are very few since all costs are known, but for the hobbyist and small time beekeeper who manage their hives in their spare time it can be difficult to arrive at satisfactory costings especially when, having arrived at sensible figures, such beekeepers have to compete with Joe Bloggs who is selling his honey at 50p per pound over the garden gate. It might be a wise move to offer him 55p per pound and buy the lot!

There are as many permutations of costing as there are beekeepers, so it would be quite impossible to give hard and fast figures, all that can be given are guidelines on which to base a realistic price at which honey should be sold.

Basically all costings have to be based on two factors. The first is depreciation on equipment, replacing lost colonies and interest based on the return if a similar sum of money was invested; the second factor is annual running costs. Once these are determined it is easy to work out a figure on which to base the selling price of honey.

For the hobbyist beekeeper who purchases five hives new from any of the suppliers, together with bees the cheapest extractor and handling equipment and who keeps his bees permanently at the bottom of the garden, the costs, based on 1988 prices would be as follows:

1 Hive, Modified National, in the flat, with 2 supers and queen excluded – hive in Western Red Cedar, collected by purchaser	£101.70
VAT @ 15%	15.26
	£116.96
6 Frame coloney of bees (VAT Zero rated)	73.00
Total	£189.96

Extracing Equipment:

1 Lightweight polythene extractor with side handle	£129.44
1 60lb Polythene honey tank with tinplate strainer	30.00
	£159.44
VAT @ 15%	23.92
Total	£183.36

Initial Outlay:

5 Hives	£584.80
5 Stocks of bees	365.00
Extracting equipment	183.36
Protective clothing, etc.	20.00
Tools	20.00
Miscellaneous	30.00
	£1,203.16

Annual Ownership costs:
Depreciation charge for all except
  bees, allowing 20 year life   £41.91
Replacement of bees (10% p.a.)   36.50
Interest on investment (10%)   120.32

                 Total   £198.73
                           =====

         Cost per hive (A)   £39.75

Annual running costs per hive:
Sugar - 20kg @ 50p/kg   £10.00
30 Jars (£17.50 gross + VAT)   4.19
Labour - 6 hrs at £2 per hour   12.00
Subscriptions (est. £10)   2.00
Insurance (1%) £12.03 for 5   2.41
Transport   -

         Cost per hive (B)   £30.60
                           =====

Total annual cost per hive (A + B)   £70.35
                           =====

Break-even price per lb, 30lbs per hive   £2.35
Break-even yield at £1 per lb.   70.35   lbs/colony
Break-even yield at £1.20 per lb.   58.63   lbs/colony
Break-even yield at £1.50 per lb.   46.90   lbs/colony

Similar costings for a 100 colony unit:-
Initial outlay:
100 hives   £11,696.00
100 colonies of bees   £7,300.00
Extractor, 24 frame radial (£578.90 + VAT)   £665.74
Straining tanks (2 x 100lb) (£136 + VAT)   £156.40
Feeders, 1 gallon (£16.20 per 10)   £162.00
Storage buckets, 30 lb (£17. per 10)   £170.00
Protective clothing (£40 per 5 years)   £8.00
Tools - including sawbench   £500.00
50 Nucleus hives (£35.87 flat + VAT each)   £2,062.53
Buildings   £5,000.00
Miscellaneous   £200.00

                 Total   £27,920.67
                           ========

Annual Ownership costs:
Depreciation charge — £1,031.03
Replacement of bees £730.00
Interest on investment £2,792.07

Total £4,553.10
=======

Cost per hive (A) £45.53

Annual running costs, per hive:
Sugar, 20 kg at £443 per tonne £8.86
Jars (£14. grs, inct VAT) £3.68
Labour - 6 hrs @ £2/hour £12.00
Subscriptions (£10) £0.10
Insurance (1%) £279.92 per 100 hives £2.80
Transport, 20p per mile, 20 visits, 60 miles each £2.4
    £240 per 100 hives

Cost per hive (B) £29.84
=====

Total annual cost per hive (A + B) £75.37

Break-even cost per lb, 31lbs per hive   £2.51
Break-even yield at £1 per lb.   75.37 lbs/colony
Break-even yield at £1.20 per lb.   62.80 lbs/colony
Break-even yield at £1.50 per lb.   50.25 lbs/colony

These figures are horrific and if nothing else they serve to show that, working on the same basis, it costs more for the semi-commercial beekeeper to produce honey than it does for the hobbyist. It is obvious that no semi-commercial or commercial beekeeper could think of buying direct from a supplier without a discount to start with, nor would he buy hives of Western Red Cedar, the same applies to nucleus hives. With luck and co-operation on all sides, domestic premises could be used for extracting and bottling honey, that would still leave the necessity of an outbuilding to act as a store cum workshop; a lock-up garage might suffice. These would bring down the initial outlay and consequently reduce depreciation and insurance. The cost of producing colonies is considerably less than the price paid to suppliers so replacement of colonies would also be less. Little could be done about specialised equipment such as extractors, straining tanks, storage buckets and feeders.

Assuming that a semi-commercial beekeeper could cut his costs on the initial outlay by 50% the costs would then be:-

Annual ownership costs:

Depreciation on all except bees		£515.52
Replacement of bees		£365.00
Interest on investment		£1,396.04
	Total	£2,276.56
	Cost per hive (A)	£22.77

Annual running costs:

Sugar		£8.86
Jars		£3.68
Labour		£12.00
Subscriptions		£0.10
Insurance		£1.40
Transport		£2.40
	Total	£28.44

Total annual cost per hive (A + B)	£51.21

Break-even cost per lb. 30lbs per hive	£1.71	
Break-even yield at £1 per lb.	51.21	lbs/colony
Break-even yield at £1.20 per lb.	42.68	lbs/colony
Break-even yield at £1.50 per lb.	34.14	lbs/colony

The conclusions to be drawn from these figures are all too obvious to need stating, and the following points should be made:-

1. No beekeeper should sell honey at less than cost price.

2. Careful siting of hives must be done to ensure maximum yields. For the semi-commerical beekeeper this means migratory beekeeping. Those beekeepers who take bees to the heather have two advantages over other beekeepers, an additional crop and the winter feed.

3. Since the loss of a swarm means the loss of the honey crop for that

season (possibly excluding heather) adequate swarm prevention measures have to be part of normal beekeeping practice.

4. In those areas where oilseed rape honey can be up to 50% of the total crop, management techniques must be geared to producing strong colonies in spring.

5. Weak and unproductive colonies cannot be tolerated so programmes aimed at disease control and queen replacement must be used to maintain colonies at maximum strength.

6. Pollination contracts at realistic prices - £20 per colony - should be sought to help reduce the costs of honey production.

7. No hive product such as wax should be wasted and outlets other than the trade should be sought, e.g. hardware shops and the making and selling beeswax polish.

Chapter 7:
Showing and Judging

Before looking at the way in which a judge goes about his task it is worth remembering that the whole business of arranging a honey show and ensuring that everything goes smoothly depends on the efficiency of the two officials, the show secretary and the judge's steward, the former because without his expertise any show would become completely chaotic, the latter, who if he failed to carry out all the small tasks immediately concerned with judging would reduce any judge to a state of exasperation.

The show secretary, having drawn up the show schedules usually in conjunction with his committee, has to despatch these to each potential exhibitor together with the show Rules and Regulations and entry forms. He has to receive the entries and entry fees, issue labels for exhibits, supervise the erection of exhibition stands, stage the entries and generally ensure that all is ready at the appointed time of judging. Not only that, he usually has the task of making the exhibition appealing and educational. He may also be responsible for any honey and other hive products which are on sale to the public and reimburse the suppliers of such products. During judging he may, if the judge so wishes, exclude the public and prohibit smoking. Show secretaries should also have available a set of honey grading glasses to British Standard 1656-1950. It is usual with most large shows to base the show rules and regulations on those used for the National Honey Show, these should be read most carefully by exhibitors and the judge.

The show steward has to ensure that the judge is left free to judge, and will ensure a supply of tea towels and water. Prior to the judging he will obtain from the show secretary a list of multiple entries so that the judge does not award first, second and third to the same exhibitor and then have to re-judge the class. He should also have previously prepared cards, one for each class, on which to list the awards in order of merit, these lists he will hand to the show secretary and upon completion of the prize cards will display then against the appropriate entry. Most judges appreciate having jar lids loosened and screws securing the tops of showcases containing frames removed. As the judge uses his tasting rods the show steward will wash and dry them, ensuring that the judge has an adequate supply at all times. After judging he will replace all exhibits in their pre-judging positions.

It is a requirement that any candidate for the Examination in Honey Judging acts as a show steward in a number of major honey shows being judged by a nationally recognised honey judge. If the judge is aware of this he usually takes the show steward into his confidence as he judges so that his expertise is passed on.

Apart from the white coat and probably white hat, the traditional wear for judges, a judge should have a set of grading glasses, tasting rods or similar devices, two empty honey jars with lids, a white card 6 inches square, a powerful torch and if he feels he needs them, a pair of sheets of polaroid film 3 to 4 inches square and a refractometer. Tasting rods can be made up taking 5 inch length of glass rod 5mm in diameter, heating the ends of each and flattening and bending them with a pair of pliers when the glass is sufficiently plastic. The white card is placed behind jars of liquid honey when checking for foreign matter, air bubbles and incipient granulation, it gives a standard background to each jar. The polaroid film is used, one as a polariser behind a jar of liquid honey, the other as an analyser in front of the jar to detect foreign matter. When the analyser is rotated so that it is at 90° to the polariser no light passes through it except that from foreign matter. The refractometer is used to give a direct reading of the moisture content of the honey being judged. The grading glasses, one light and one dark determine the limits of honey classed as medium, any entry which is lighter than the lighter glass should be in the light liquid class, anything darker than the darker glass should be in the dark liquid glass.

In small shows medium and dark classes are often combined.

Fig. 24 The tools of the trade

Fig. 25 A polariscope used for judging

Fig. 26 Plans for polariscope

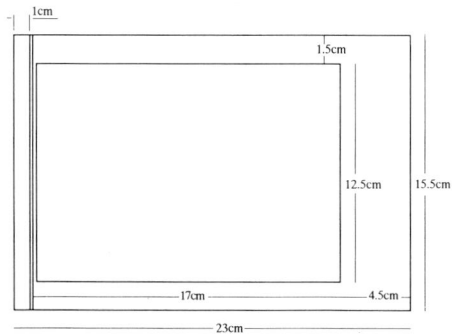

Top of lamp housing from below

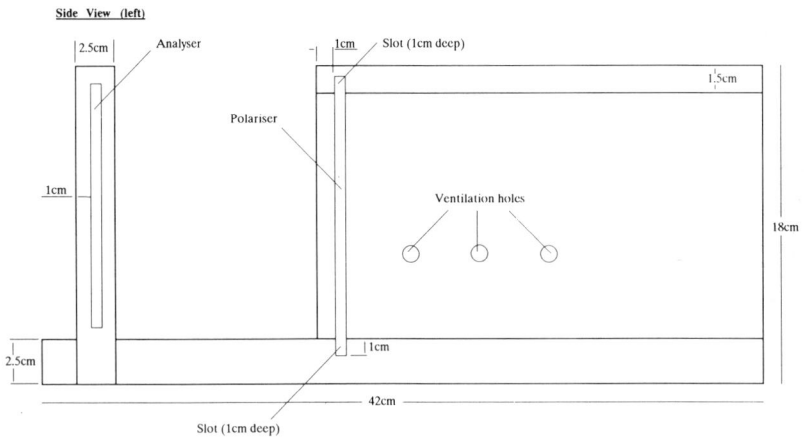

Side View (left)

56

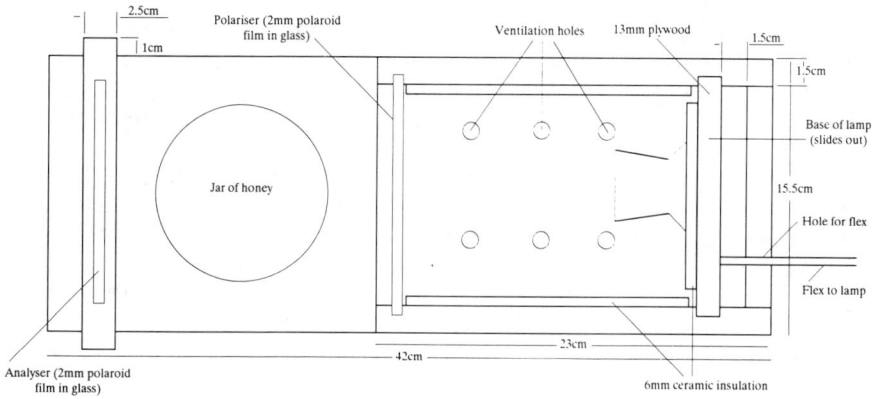

2.5cm

Polariser (2mm polaroid
film in glass)

1cm

Ventilation holes   13mm plywood

1.5cm

1.5cm

Base of lamp
(slides out)

15.5cm

Hole for flex

Jar of honey

Flex to lamp

Analyser (2mm polaroid
film in glass)

23cm

42cm

6mm ceramic insulation

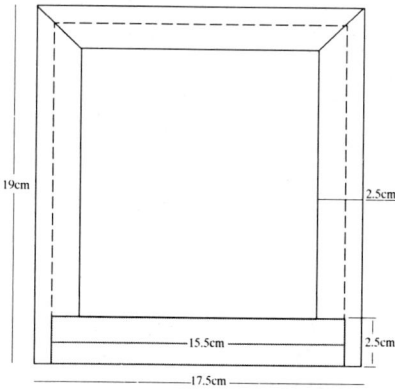

19cm

2.5cm

15.5cm

2.5cm

17.5cm

Front View

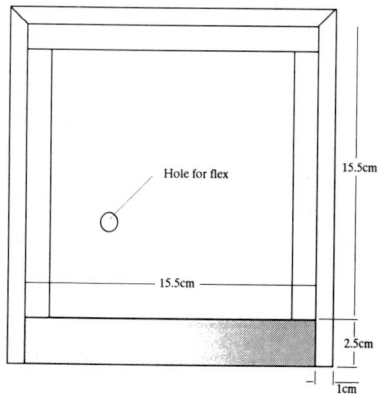

Hole for flex

15.5cm

15.5cm

2.5cm

1cm

Back View

57

Liquid Classes

Before the judging starts the judge must study the show schedule and may check the colour of entries in each class to ensure that all are in the correct class and, if needs be, reclassify those which are close to the borderline between light and medium and medium and dark. Even with very great care on the evening prior to judging when the light may be very dim, some entries may be staged in the wrong class.

Most judges use the same procedure after checking that all entries in a class conform with the regulations for that class, apart from colour this entails checking that each pair of jars in an entry match in all respects and have the same lids, plain lacquered and unpolished. Coloured lids are not allowed. A judge may not look with favour on any jar which has obvious imperfections in the glass on the grounds that if the exhibitor cannot take sufficient care over his choice of jar such an entry does not deserve much consideration.

Any entries which the judge considers unworthy of further attention are usually moved to the back of the show bench or stood one on top of the other; as further entries are eliminated they are treated in the same manner.

The judge will next examine each jar for cleanliness and clarity, any entry which contains foreign matter, air bubbles or incipient granulation is eliminated at this stage. Any of these faults can be detected either by shining a powerful torch through the jar, standing the jar in front of the white card or using the polaroid srips. In some cases where the light is dim inside the exhibition building the judge may need to take some exhibits outside in order to examine them thoroughly.

By now the judge has examined them sufficiently from the outside and is left with those he considers worthy of further examination. One jar from each pair is taken and the previously loosened lid is carefully removed and the aroma tested; the inside of the lid, the surface of the honey and the inside of the jar above the level of the honey are all checked for cleanliness and lack of scum. Where wads are used inside the lids the judge should be on his guard in case the exhibitor has used an enhancer to improve the aroma; this is almost impossible to do with 'flowed-in' plastic. Apart from testing for aroma all the work a judge has done so far has been objective; testing for aroma and flavour are purely subjective and the results depend to a very large extent on the judge's personal preference, although there appears to be a large measure of agreement. It is not unknown for an entry to obtain a first prize at the National Honey Show and no award three weeks later at a country show when in both cases the judges have been nationally recognised and holders of the judge's certificate.

Flavour and density are tasted almost simultaneously, the judge dips the tasting rod into the honey and on withdrawal watches how long the surface of the honey takes to level off. The longer it takes the denser the honey and consequently the less the water content. Some judges test for the latter using a refractometer. The drop of honey is then wiped off on a finger, often the back of the forefinger, for tasting. Continuous tasting not only makes the finger sticky but also dulls the sense of taste, so frequent cleaning of finger and occasional nibble of something refreshing to revitalise the taste buds are necessary. Many judges eat a small piece of a cooking apple. It is not a bad policy if judges, having arranged the exhibits in order of merit, tests the second jar of each entry, in reverse order, having first cleansed his mouth. This ensures fairness, particularly if there is little difference between them.

There should be no difficulty in detecting foreign matter, air bubbles and early stages of granulation in light and medium honey, but dark classes can present problems when the honey is so dark as to be opaque even with a very strong light. Although early stages of granulation are easily detected when the honey is being tasted and air bubbles will show themselves on the surface of the honey it may be almost impossible to detect foreign matter and except where there are glaring examples, all entries should be treated on the same basis for fairness.

Although there are no separate classes for honeydew honey in the British Isles because pure honeydew honey is uncommon, it is not unusual for a proprotion of such honey to be present in some of the darker honeys. Most judges, if they recognise the presence of honey-dew honey in dark honey, will treat such entries on their merits.

Creamed or Granulated Honey

In many ways this is one of the easiest classes to judge because such honey is difficult to have in first class condition on the day of the show. The two classes are not separated here because the difference between creamed and granulated honey is usually a matter of crystal size and firmness. In many shows both types are exhibited in the same class; a finely naturally granulated oilseed rape honey can be almost indistinguishable from a creamed honey containing a lot of oilseed rape honey. Whereas naturally granulated honey is usually fairly hard and may be difficult to spread, creamed honey should have the consistancy of butter or margarine and spread easily.

Both naturally granulated and creamed honey are lighter than the same honey when liquid due to the glucose crystals being white, this makes the detection of foreign matter, at the bottom of the jar and close to the sides, very easy. The judge having checked the jars and lids for uniformity and conformity will invert the jar being judged and inspect the bottom, rejecting any with foreign matter, also rejected will be any that show signs of frosting. Having removed the lid the judge will check it for cleanliness, inspect the surface of the honey for particles of wax and similar matter, also the presence of frosting. He will then tip the jar on its side and reject any that show signs of movement. The surface of the honey should be smooth and dry although slight moistness is not usually penalised. As with liquid honey, creamed or granulated honey will be tested for aroma when any fermentation will reveal itself by the smell. Both types of honey ferment more readily than liquid honey and the presence of a layer of wet honey or bubbling on the surface will also indicate fermentation later verified by tasting. Having eliminated all exhibits except those he wishes to taste the judge then removes a small piece of honey, close to the glass, with his tasting rod and tests it for granulation and flavour. Apart from looking for the finest flavour the judge will also give preference to the honey with the finest granulation. Those honeys which have large hard crystals and a very firm texture are not usually among the highest awards.

Heather Honey

For show purposes heather honey should not be granulated, but in its thixotropic state. When pressed out it should be strained so that no foreign matter is present. Because even in its liquid state heather honey is very viscous, air bubbles cannot rise to the top; the purity of heather honey is indicated by the bubbles present. In very viscous honey, that having the highest density, large air bubbles may be present. Air bubbles should be evenly distributed throughout the honey and they should not form a layer on the surface. As with creamed or granulated honey the jar should be turned on its side, there should be no movement of the honey. The thixotrophy and density of the honey is also tested by moving the end of the stirring rod rapidly to and fro to create a furrow on the surface of the honey and watching how long it takes for the surface to flatten out. The honey should then be tested for flavour as heated heather honey easily caramelises.

Comb Honey

Much of what a judge is looking for in comb honey be it section, frame or cut comb has already been dealt with. If comb honey is being exhibited in sections or frames the cleanliness of the woodwork is most important and the judge will give preference to those having the cleanest woodwork. As mentioned in an earlier chapter all cells should be sealed and the cappings level and even. Preference is given to the cappings on each side could be removed easily and with one stroke. No cell should contain pollen or granulated honey, both are detected by showing a light through the comb and observing from the other side. Cells containing pollen will show up dark, those containing granulated honey will have a 'dead' appearance. There should be no weeping from the cells, this indicates damage to the cappings. Exhibits showing excessive tunnelling by the larvae of braula coeca usually incur penalties, some judges may use a magnifying glass to detect these.

Chunk Honey

Not many honey shows, except the National and large regional shows, have sections for chunk honey. For show purposes the amount of comb present should be about 50 per cent. When judging, the criteria used are those one would use when judging liquid and comb honey. The liquid honey should be light in colour so as not to obscure the comb. To arrive at a fair judgement the exhibits in this class are best judged on a points system.

The points system of judging has not been mentioned previously as most judges use a process of elimination based on a set of standards he carries around in his head, and this is the system on which the foregoing remarks on judging honey have been based. In theory the points system, so adequately described in Beekeeping Techniques by A.S.C. Deans, is the ideal one to use and probably works very well with small classes, with large classes, however, because it is tedious and the judge has a limited time in which to make his decision, the points system is almost impractical.

Composite Classes

These are often the bane of many judges as they can vary from one show to another and because of their composition they can be difficult to judge. They vary from jars of honey of different colours to those with creamed or granulated honey; they may include comb honey in one form or another and wax, either plain or fancy moulded, with the odd candle or bottle of mead thrown in for good measure. It must be said in their defence that entries in such classes are often very attractive and much admired by the general public. It is almost impossible to do justice to entries in this class unless each particular part of the exhibit is judged using a points system.

61

Becoming a Honey Judge

There are judges and judges. But to be able to judge at the National Honey Show of any country in the British Isles or at a major county show, it is necessary to hold the Honey Judge Certificate of that country. Having said that, a judge has to start somewhere, it might be at a local village flower show where a local, well known beekeeper might be used and everyone would be quite happy. Somewhere along the line the organisers of a honey show will have to justify their choice of judge. The possession of the National Diploma in Beekeeping, the highest beekeeping qualification in the British Isles and recognised as such in many English speaking countries in the world does not qualify the holder to judge, but at least it does show that the holder has had to pass those sections of the examination that apply to honey and judging. Candidates are required to know how honeybee products are prepared for the showbench, and judged; they are also required to know the legislation which is pertinent to all aspects of marketing honey bee products, together with extensive knowledge of the composition of honey and other aspects of the chemistry of honey – beeswax. They will be examined practically on the various aspects of judging honeybee products including deterioration and faults. At least if it doesn't qualify for the National, its not a bad start!

The three national beekeeping associations, B.B.K.A. to which associations in England are affiliated, S.B.K.A. which covers Scotland and F.I.B.K.A. to which association in Ireland including the province of Ulster are affiliated, all award a Honey Judge Certificate to successful candidates.

The B.B.K.A. has recently introduced a new qualification for judging, the Associate Judge Certificate for which candidates need only hold the B.B.K.A. Preliminary Certificate; they must have acted as a honey steward in at least three major shows and can show evidence of their ability as exhibitors. This certificate will qualify the holders to judge in all but major honey shows. In future any candidate for the Honey Judge Certificate will have had to have passed the Associate Judge examination.

The requirements for examination for the Honey Judge Certificate of any national association are almost identical, the Practical, Intermediate or Senior Certificates of the B.B.K.A., the Bee Masters Certificate of the S.B.K.A. and the Intermediate or Senior Certificate of the F.I.B.K.A. and to have acted as a judge's steward at two major shows. They must also produce evidence of having been a successful exhibitor at major honey shows.

The syllabus ranges widely over knowledge of all forms of bee products and appliances as generally shown, common faults and fakes, usual show practices and rules together with the methods commonly used for making awards.

The examination consists of an oral and a practical test usually held at the national honey show of the country concerned when the candidate will be asked to judge certain classes or equipment and give comments and reasons for his decision. He is also examined on his own samples of honey, both liquid and granulated, beeswax, mead and comb honey as prepared for exhibition, but is not usually penalised for faults in his exhibits due to circumstances beyond his control.

Some associations run courses to help prepare candidates for examination in judging.

Details of syllabuses and other information regarding these examinations can be obtained from:

B.B.K.A. Mr. A. Barber,
Secretary, Education and Examination Board,
Charnwood,
Beechfields,
Barlaston,
Stock on Trent, ST12 9AP.

S.B.K.A. Mr. William, A. MacKenzie,
General Secretary,
"Crougie",
9 Glenholme Avenue,
Dyce,
Aberdeen, AB2 0FF.

F.I.B.K.A. Mr. Peter Whyte,
Educational Convenor,
Laska,
Riverstown,
Birr,
Co. Offaly,
Ireland.

BIBLIOGRAPHY

A.S.C. Deans *Beekeeping Techniques*, Oliver & Boyd.

Eve Crane *Honey – A Comprehensive Survey*, I.B.R.A.

Crane, Walker & Day *Directory of Important World Honey Sources.* I.B.R.A.

John B. Free *Bees and Mankind,* George Allen & Unwin.

Ted Hooper *The Guide to Bees & Honey*, Blandford Press.

Morse & Hooper *The Illustrated Encyclopedia of Beekeeping*, Alphabooks.

A. I. Root *A. B. C. and X.Y.Z. of Bee Culture*, The A.I. Root Co.

M.A.F.F. Bulletin No. 134, *Honey From Hive to Market*, H.M.S.O.

J.T.W. Scruby *Articles on judging honey and hive products.* Bee Craft. September, October, November 1988. *Bee Craft Ltd.*